*Congrats on your graduation & summa cum laude award! May this book inspire you more about mathematics!*

TO: Anna Benbrook,

*This book is my memoir. It is certainly presumptuous of me to even think that you might want to read such a tome. But, without my either informing you of this or giving you a copy, as I am, you might never know it existed. This is a gift to you because I value you as a friend, so much so that if you wrote such a book, I would certainly be disappointed not to at least have the choice to read it.*

*I know that mathematics can conjure up fears and reservations in many people, but this is not a mathematics textbook. In fact, you can easily read around the mathematics for the many stories that permeate the book. For example, look for the topic about Elaine in Chapter 1, the humorous side of mathematics in Chapter 4, and the stories of people who touched my life in Chapter 5.*

*This book is not an autobiography, but it is autobiographical. It is not a math textbook; but it does contain lots of mathematics. It is not a book on the history of mathematics, but it does contain some history. It is not an education book, though it does contain my theories on math education. It is not a book on applications, but most of my favorites are discussed. It is not a book on philosophy, but my spirituality will be on display throughout. In short, it is* **One Man's Journey Through Mathematics**. *It will be my joy if you find it meaningful in some way!*

*Some of you are aware that I have also been writing a book entitled* **Mathematics and the Spirit**, *which attempts to integrate topics from mathematics and the Christian faith. That book will follow soon.*

5/12/05

*Marvin L. Bittinger*

**MARVIN L. BITTINGER**
3011 Whispering Trail • Carmel, IN 46033
(317) 844-7947 • Fax: (317) 846-2292 • e-mail: exponent@aol.com

# One Man's Journey through Mathematics

**MARVIN L. BITTINGER**

PEARSON

Addison
Wesley

Boston   San Francisco   New York
London   Toronto   Sydney   Tokyo   Singapore   Madrid
Mexico City   Munich   Paris   Cape Town   Hong Kong   Montreal

**PHOTO CREDITS**

Cover: Karen E. Bittinger

p. 20 Beth Anderson, p. 40 Karen E. Bittinger, p. 45 Michael St. André, p. 48 Michael St. André (top, middle, and bottom), p. 49 Michael St. André, p. 56 (bottom) Bettmann/Corbis, p. 59 Courtesy: Donruss Trading Cards, p. 69 Bettmann/Corbis, p. 70 AP/Wide World Photos, p. 71 (top and bottom) Bettmann/Corbis, p. 75 Bettmann/Corbis, p. 76 Bettmann/Corbis, p. 78 (left) Reuters NewMedia Inc./Corbis, (right) Duomo/Corbis, p. 79 Bettmann/Corbis, p. 80 BSF/NewSport/Corbis, p. 91 Neal Preston/Corbis, p. 92 Troy Wayrynen/NewSport/Corbis, p. 100 (top) Bettmann/Corbis, p. 100 (bottom) Michael St. André, p. 102 Columbia University Archives and Columbiana Library, p. 103 University of St Andrews MacTutor History of Mathematics archive, p. 104 University of St Andrews MacTutor History of Mathematics archive, p. 105 University of St Andrews MacTutor History of Mathematics archive, p. 106 Ed Kashi/Corbis, p. 117 1961 *Aurora*, Manchester College, p. 121 1961 *Aurora*, Manchester College, p. 127 1961 *Aurora*, Manchester College, p. 129 Air Force Military Museum, p. 130 US Air Force, Staff Sgt. Andy Dunaway.

"The Algorithms of Sport" by Dennis Bergendorf as it appears on pp. 50–54 is reprinted with permission from *Bowlers Journal International*.

Arby's® and New Hat Design is a Registered Trademark of Arby's IP Holder Trust.

PEANUTS reprinted by permission of United Feature Syndicate, Inc.

Bittinger *Introductory Algebra*, 9/e pp. 159, 185, 308–09, 311, 511–12, 569, and 616, ©2003 Pearson Education, Inc. Reprinted by permission of Pearson Education, Inc. Publishing as Addison-Wesley.

Bittinger, *Basic Mathematics*, p. 136 ©2003 Pearson Education, Inc. Reprinted by permission of Pearson Education, Inc. Publishing as Addison-Wesley.

Many of the designations used by manufacturers and sellers to distinguish their products are claimed as trademarks. Where those designations appear in this book, and Addison-Wesley was aware of a trademark claim, the designations have been printed in initial caps or all caps.

**Library of Congress Cataloging-in-Publication Data**

Bittinger, Marvin L.
    One man's journey through mathematics/ Marvin L. Bittinger.
        p. cm.
Includes bibliographical references.
ISBN: 0-321-24150-9
    1. Mathematics—Miscellanea. 1.Title.

QA99.B57 2005
510—dc22                                                                2004052879

# Foreword

MARVIN L. BITTINGER is the author of more than 170 college mathematics textbooks published by Addison-Wesley Publishing Company. By our estimate, over his 34-year career Marv's books have sold more than 11 million copies in the United States and around the world. He is likely the most successful higher-education mathematics author of all time.

Marv's three-book series on developmental mathematics created something entirely new in mathematics education. Most mathematics authors improve on an existing subject or method, but Marv actually created an entirely new approach to remedial college teaching with his series of paperback books by incorporating margin exercises, behavioral objectives, and a unique book design. Professors could virtually pass the teaching responsibility from themselves to the students through these innovative textbooks.

Marv's textbooks are renowned for their clarity, accuracy, and organization. There are thousands of teachers who initially learned their mathematics from one of Marv's texts. Many of these teachers are currently using his books to inspire a new generation. Most important, Marv's texts have enabled millions of students who may have struggled to learn mathematics previously to succeed in mathematics.

—GREG TOBIN
Vice President, Publisher,
Mathematics and Statistics
Addison-Wesley Publishing Company

*This is a book of the kind and wacky genius one is lucky enough to encounter but a few occasions during his or her lifetime.*

—M. SCOTT PECK

# Contents

**CHAPTER 7**

## ON COLLEGE MATHEMATICS EDUCATION 175

# How I Became
# a Mathematician

**INTRODUCTION**   This book is not an autobiography, but it is autobiographical; it is not a math textbook, but it does contain lots of mathematics; it is not a book on the history of mathematics, but it does contain some history; it is not an education book, though it does contain my theories on math education; it is not a book on applications, but most of my favorites are discussed; it is not a book on philosophy, but my spirituality will be on display throughout. In short, it is *One Man's Journey through Mathematics.* It is my prayer that you will find it meaningful!

I'm 62 years old as I write this book. Thanks to my publisher, Greg Tobin, at Addison-Wesley Publishing Company, I am being allowed to create what I feel is a "crowning" event in my mathematics textbook writing career.

I sometimes feel self-centered using the word *I*, but I can't share my life, career, and heart without its use. I write this book acknowledging that it might be construed by some to be an ego trip. I'll take that criticism on the chin in the hope that many will derive some inspiration from my joy of being able to creatively present mathematics in the classroom and on the printed page for more than 34 years.

I know that in a mathematics textbook, one should deal with the subject matter and relinquish one's philosophy and religious viewpoints out of respect to students and instructors of all religious and ethnic backgrounds. I do abide by that credo in my math textbooks, but this book is not a textbook; it is a unique endeavor that reveals my profession, my hobbies, and the spiritual side of my life. In this book I take the liberty of expressing my philosophy on life, education, and faith as I see fit.

I think every mathematician has a streak of philosopher in him. You can see this in the lives and works of people like René Descartes (1596–1650), Blaise Pascal (1623–1662), Isaac Newton (1642–1727), Alfred North Whitehead (1861–1947), and Bertrand Russell (1872–1970). Indeed, I hold the strong belief that all scientists, including mathematicians, should return to a union of faith and reason in their daily and professional lives. Space scientist Werner von Braun once said, "[It is] as difficult to understand a scientist

*The object of mathematics is
the honor of the human spirit.*
—CARL JACOBI, professional
mathematician, 1804–1851,
Potsdam, Prussia

who does not acknowledge the presence [of God] . . . as it is to comprehend a theologian who would deny the advances of science." More will be said about this premise toward end of the book.

I am therefore very grateful to my publisher for allowing me to combine my profession, my philosophy, and my heart in this book. While I get to say what I think and feel, I do so with all respect to the diversity of my audience. At the same time I ask for the same respect. I want those of you who think that math textbook authors are not real people to see some of the struggles and joys involved in my work and perhaps see the personality of an author who seems far away and detached when they study a math textbook.

To read this book, all you need by way of background is some knowledge of college algebra and trigonometry and sufficient interest in the ways of mathematics, mathematics education, and the authors of mathematics textbooks.

## EARLY BACKGROUND

Born August 9, 1941, I was raised in a middle-class family in Akron, Ohio, by my grandparents because my mother, Ersle I. Bittinger, fell out of the hospital window 10 days after my birth. Only in recent years have I surmised that this may have been a suicide or accident due to postpartum depression. I was told that my mother had some kind of omen about death at childbirth and she had asked my grandmother, Bessie Gosnell, to raise me in the event of her death. That request together with the fact that my father, Sterl E. Bittinger, was about to enter World War II, led to my being raised by my grandparents.

My grandmother, matriarch of our family, was the dominant person in my upbringing. I shall thank her always for raising me in a Christian home and teaching the value of a college education to her children and to me. There was never an issue of whether I was going to college—it was only where and how. I made $5 a week delivering groceries to a nursing home for children who had polio. I remember being taught to save $3.00, spend $1.50, and tithe $0.50 to the church. Later, I delivered newspapers for the *Akron Beacon Journal* and used my income in the same proportions.

I was handicapped with what used to be called a "lazy" eye and received some vocational rehabilitation money from the state of Ohio. That money, help from my father, and the money I saved from my grocery and newspaper delivery routes and college summer painting jobs funded my education so I was not in debt when I finished. But, then, my entire college education only cost about $3300.

I wish I could weave a glamorous story of my passion for math as a child. Yes, it was a favorite subject, but never in grades K–12 did I even slightly consider being a math major, or a professor, let alone a textbook author.

The closest connection to writing was the attention to detail I enjoyed building model airplanes. I used to spend winters in solitude in a cold, damp converted coal bin in our basement, listening to music and constructing those planes, only to face the disappointment in the spring of crashing them or losing them as they soared away.

There are two "math events" that I do recall vividly. The first in my youth occurred when I was walking home from school one day. I remember almost exactly where I was on Eastlawn Ave. in Akron, Ohio. I worked out an informal proof in my head that there could not be a largest number. The proof went something like the following:

> *Suppose there is a largest number. Then add 1 to that number and you have a larger number. But that contradicts that we just said there was a largest number. Thus, there is no largest number.*

Pure mathematicians: Hold your letters critiquing this proof. I might have been age 10 at the time.

The second math event occurred in the 11th grade at East High School in Akron. I received a D as my 6-week grade in 11th-grade algebra. I know now that my teacher, Dorothy Leffler, was making an object lesson of this grade because she saw how fast I was handing in my tests. She wanted me to go over my work carefully and win the race of getting an A next time, instead of winning the race to turn in my test.

Now, if you check my high school transcript, you will not find this D; only semester grades appear. Yet, it may have foreshadowed my interest in mathematics education. At least it shows that math authors can struggle and be just as human as students.

## HIGHER EDUCATION

My two maternal uncles, Wilbur and Rex Gosnell, 18 years and 11 years my senior, respectively, are very intelligent men who went to college at the urging of their mother (my grandmother). Wilbur was an engineer at Goodyear Tire & Rubber in Akron and Rex received his Ph.D. in chemistry from Purdue University, working first for Union Carbide and then operating his own chemical research business in San Diego. Rex was truly my education hero.

I knew I was going to college, but where and what my major would be were uncertain. We attended Eastwood Church of the Brethren and most of the youth at that church attended our church college, Manchester College in Indiana. My mother went to Manchester College, as did my two uncles. At the recommendation of my uncles, who had chosen careers in applied science, I decided to enter Manchester College with the intent of being an engineer.

Students who are reading this book will probably identify with the next stage of my education. I took chemistry as a freshman and had to work hard to get Bs. When I was a sophomore, I took physics and struggled for Cs. The thought soon came to mind, "How can you be an engineer with these chemistry and physics grades?"

My math classes went far better—I saw two different worlds. It was truly a quirk of fate, or a "God thing," or an epiphany, when I ended up in calculus as a freshman. I began by taking an algebra-trigonometry class with Professor John K. Baumgart. One of the first topics we considered was determinants and the solving of systems of equations. In all honesty, I struggled with those determinants. I understood the theory but, without a calculator, made computational mistakes. (I still hate to balance my checkbook, leaving it to my wife.) But I was always one to ask questions and was always intrigued by the mysterious word *calculus*. So, after class one day, I asked Professor Baumgart what calculus was all about. I wanted nothing more than a brief explanation, which he so graciously gave me. After the explanation, he said, "Why don't you go down and try the calculus class?" Someday, in another life, I want to ask Professor Baumgart what prompted him to make that offer. I swallowed a quick gulp and thought, "How can I do calculus when I am struggling with determinants?" Nevertheless, I accepted and entered Professor David L. Neuhouser's class the next day.

### Math Power Experiences

There will be many times in this book where I relate to you my joy regarding a personal epiphany in my life. Inevitably such an epiphany involved mathematics. Thus, I will call each each event a Math Power Experience and label it MPX . Such events touched me so much that in order to share them with you I might make a suggestion of some kind of music or a movie to accompany this experience. Were space available, the events would include photos from my beloved Arches and Canyonlands National Parks. Please indulge this practice—I am sharing my heart, but *you don't have to play the music or rent the movie.*

You often hear stories about how people rise to a challenge. Being in that calculus class as a freshman, not often done in those days, was a tremendous boost to my self-esteem. I worked very hard and was proud of my situation. I got a B that first quarter, then As the next two quarters, and those As continued the rest of my undergraduate math courses. I would say that Professor Baumgart's suggestion was truly a math power experience.

MPX    MUSIC: "The Welcoming," Original music from the motion picture sound track, *The Big Country*, 1958, United Artists Records, Inc.

In my sophomore year at Manchester College, I took advanced calculus with Professor Baumgart. It was a hard course, yet I made an A though I didn't feel I had earned it.

There was soon another MPX in my life that I remember to this day. I was walking beside the Manchester College library when I heard a still, small voice say, "Marv, what do you like best in college?" The answer was *math*. The next question was, "What do you get the best grades in?" The answer again was *math*. "Then why don't you major in *math*?" I know now that voice came from God—it was indeed another MPX in my life.

MUSIC: "Life Is a Dream," from the sound track to the movie, *Star Trek: The Final Frontier*, copyright © 1989 CBS Records, Manufactured by Epic Records.

The next quarter I dropped physics and changed to classes directed toward a degree in mathematics education. This event, happening at age 19, so affected my life that recently I shared it with my son Chris when we visited Manchester College as he carried out his work for Campus Crusade for Christ.

## Moving On to Mathematics Education

Now, 42 years and 170 math textbooks later, I am often asked the question, "How did you happen to go into math education?" The most succinct answer is that it was because I had such a remarkable string of wonderful math teachers. I can name them all the way from 7th grade through graduate school:

Betty Mae Burke, 7th grade, Ellet Junior High school
Dorothy Viola, 8th grade, Goodyear Junior High school
Martha Miller, 9th grade, Goodyear Junior High school
Dorothy Leffler, East High School
John K. Baumgart, Manchester College
David L. Neuhouser, Manchester College
Norman Levine, Ohio State University
Angelo Margaris, Ohio State University
Christoph J. Neugebauer, Purdue University
Mervin L. Keedy, Purdue University
Robert A. Osterle, Purdue University

I have always been a careful observer of my teachers, especially my math teachers. When you are fortunate enough to have such a string of great role models, they inspire your interest in the entire field. I'll explain even more about some of these teachers and mathematics education in later chapters.

## Becoming a Math Textbook Author

I think I have told the story about how I became a math textbook author about a thousand times. As a student, I had the utmost respect for my math books. I treasured them so much that I never sold them as used books. I still have Peterson's *Elements of Calculus* (cost was $4.60), Rainville's *Elementary Differential Equations*, and Finkbeiner's *Matrices and Linear Transformations*. Those math authors were a marvel to me because I wondered how anyone could create anything more elegant than these math books.

When I was at Purdue University, 1965–1968, working on my doctorate in mathematics education, I took a course in advanced geometry with Professor Mervin L. Keedy, another  in my life that would symbolize the American dream. I was in his office one day asking about the course. I knew he was a very successful author at that time, having written many successful books during the modern math movement of the 60s and 70s. As I was leaving his office one day with my body against the door, I summoned the courage to ask, "Say, I have always been interested in the writing of math textbooks—how does one get started?" His response was, "The best way is to associate yourself with someone who has already written. By the way, I have a manuscript over here. It is for a programmed book on trigonometry. I have had three or four people work on it, but none of them had the desire to finish it. Would you like to give it a try?" Talk about an MPX ! It was like throwing raw meat to a wolf. I seized that opportunity with all the vigor I could manage and wrote that book while finishing my Ph.D. thesis. I got my degree in August of 1968, and that book was published in the spring of 1969—an abnormally short amount of time to have a publication following graduation.

**MPX** MUSIC: "Take Us Out," from the sound track to the movie *Rudy*, copyright © 1992 TriStar Pictures, Inc. Music composed and conducted by Jerry Goldmsith. MOVIE: *Rudy*, 1993, starring Sean Astin and Ned Beatty in the story of a young man from a blue-collar home who plays football at Notre Dame, not for fame but to fulfill the dream of his heart.

That book, *Trigonometry: A Programmed Approach,* published by Holt, Rinehart & Winston, was not well received, nor was the next book, *Mathematics: A Modern Approach*, published by Addison-Wesley, particularly successful. But, during the writing of that book our editor, Charles D. "Chuck" Taylor, signed us to do a paperback series on developmental college mathematics, called the "The Trilogy":

*Chuck went on to become a very successful writer of adventure novels in the Tom Clancy genre. Some of his best-sellers are* Show of Force, Boomer, Sightings, *and most recently* Igniter *about an arson investigator in the Boston Fire Department.*

1. *Arithmetic,* now called *Basic Mathematics,* consists of the essentials of 7th- and 8th-grade arithmetic—whole numbers, fractional notation, decimal notation, percent notation, formulas of geometry, and basics of algebra.
2. *Introductory Algebra* covers 9th-grade algebra—real numbers, equation solving, and up to quadratic equations.
3. *Intermediate Algebra* is the bare bones of 11th-grade algebra—equation solving, systems of equations and solving rational equations, ending with exponential and logarithmic functions and conic sections.

Some details on creating the Triology follow. More detail is in Chapter 6.

For those of you not familiar with the words *developmental mathematics,* they used to be called *remedial mathematics* in college circles, but that term was replaced to avoid the stigma for students who had to take such courses. This series was the first ever in paperback. Knowing we needed to learn more about the dilemma instructors faced with this remedial instruction, we did extensive traveling to community colleges throughout the United States.

Knowledge gained from such travels, together with the use of margin exercises, careful sequencing of the learning, the use of behavioral objectives, and a special book design led to an entirely different kind of instructional package.

The production department at Addison-Wesley came up with a page design that incorporated a larger-than-usual margin to the side of the main instructional text. It was my task to determine how to use that margin effectively. I remember sitting at my desk and sliding a piece of yellow paper sideways into my typewriter so I could type the text and then the margin material all on the same page. I conceived of two approaches to utilize the margin. The first was to have the student bounce back and forth between the margin and the text, doing exercises to discover certain mathematical properties. The second approach came from my experience writing programmed learning. The student would study one third to one half a page, then stop and do margin exercises to reinforce the learning. I bounced both ideas off my coauthor, Mike Keedy, and we agreed to go with the latter.

An old adage in mathematics education says, "You can't learn math without a paper and pencil in hand." Having students stop and do margin exercises brought the adage to life. In the process, the readability of the mathematics was greatly enhanced because the student became involved in the learning as he proceeded. The book became "user friendly." The instructor used less effort to convey the math to students. Ironically, some instructors even complained that they felt less needed.

Innovative notions in math education are slow to catch on. A determined Addison-Wesley delivered the message of these books by publishing a second edition that saw a steady, but not overwhelming, response. By the third edition, that message arrived and the book's sales began to sky-rocket. I was so proud of how well students were learning. It was tremendously satisfying to build on a concept you could hold up as a finished product, and even more satisfying that so many students learned from it successfully.

I guess if imitation is the surest sign of flattery, then we have been flattered. The innovations in the Trilogy continued to work well because many other authors copied them.

When the publisher saw how successful our concepts and writing had become, they approached us about more products—algebra and trigonometry, calculus, precalculus, hardback algebras, and so on. Now, in 2003, I have just finished the 9th edition of the developmental books and the 165th math textbook in all, if you count the revisions. I have been truly blessed with a special talent, and I am truly thankful to God for that blessing.

If I were to write one sentence to describe my writing, it would be that I write math books for ordinary people, somehow being able to help people understand math who previously had difficulties. I struggle with the term *ordinary* and always try to keep in mind the following statement by one of my favorite authors, C. S. Lewis:

> *There are no ordinary people. You have never talked to a mere mortal. Nations, cultures, arts, and civilizations—these are mortal, and their life is to ours as the life of a gnat. But it is immortals whom we joke with, work with, marry, snub, and exploit—immortal horrors or everlasting splendors.*

This quote is from a sermon Lewis preached at the Oxford University Church of St. Mary on June 8, 1941, entitled "The Weight of Glory." It is in the book *The Weight of Glory* and *Other Addresses*, republished in 2001 by Harper-Collins.

More details on the creation of the Trilogy are given in Chapter 6.

# The Joy
# of Applications

# 2

I must admit that I was drawn to math because of the satisfaction of working with its logic and symbolism. It was like working puzzles to solve an equation with a string of steps or to prove an identity in trigonometry with a sequence of manipulations. I agree with many people who simply enjoy the fact that you get a final, absolute answer. There are no ifs, ands, or buts in math—there is just an answer.

As a student, I did not particularly enjoy "word" problems, but I did them successfully and occasionally I really got satisfaction in solving some of them. Over the years of my writing, I have come to thoroughly enjoy finding applications of math to the real world. Most of the problems came from my daily experiences. It was a real joy to incorporate them in my writing as much as possible. In this chapter and other parts of this book I share with you my favorites. They have evolved from either my everyday experiences or my writing.

I'm often asked how you revise a math textbook. One answer lies in the changing trends in mathematics education. I've seen many changes in 30 years, from the metric system, to audiotapes, videotapes, digital videos, two-color books, four-color books, and use of the computer and the Internet. Yet the greatest impact is the use of modern applications that the student might encounter in everyday life. In every revision we update the applications, as well as offer new ones in keeping with the changes in society and the evolving college student. Why? Students often ask the question, "What is this good for?" or "What do you do with this?" The answer lies in making the mathematics more "real" by showing these applications. In all my authoring experience, achieving this task is the *most difficult part of writing.* I'm like a cat on the prowl looking for applications of math in the newspapers, magazines, TV, movies, the bowling center, the baseball stadium, and in my Christian faith. Say "*x*-factor" and my ears perk up. Show me a graph and I look at its shape to determine if I can fit it to an elementary function. Roll a bowling ball and I see a specific path down the lane. Show me a movie with a math student or professor in it and my ears turn up for the math. Take me to a baseball game and I see "Tale of the Tape" on the scoreboard and I want to know how they compute the home run distance. Take me to Wendy's Hamburgers and I want to know how they could fix hamburgers 256 ways. While the task of creating meaningful applications is the most difficult in writing, it can also be the most satisfying.

I think the perfect application has to satisfy three criteria, which I abbreviate with the acronym, **IRU:**

**I   interesting** to the student,

**R   relevant** to the student, meaning it fits into his or her hobbies, interests, and major,

**U   understandable** to the student, meaning the student must be able to understand the application, that is, have enough math knowledge to follow its development.

Now finding such an application for one student is formidable, let alone an entire class, or all the students around the world who might be studying with this book. If it satisfies U and one of the other two, then I might put it in a book; satisfying all three is close to impossible. Let me give you an example.

**THE BIRTHDAY PROBLEM:** Of $n$ people in a group, what is the probability that at least 2 of them have the same birthday (day and month, but not necessarily the same year)?

For 30 people in a room, the probability that 2 of them have the same birthday is about 0.706, which is unexpectedly quite high. For 41 people it is about 0.903, and for 100 people it is about 0.9999997.

I have a dear friend, Joe Graves, who loves philosophy and mathematics and runs his own computer software consulting company, but he has not had extensive training in math. He thought this problem was so intriguing (I & R) that he tried to solve it, but the math theory was much too difficult for him ($\sim$U), which means I was able to meet two out of three of my criteria. Most people can understand the statement of the problem and the result, but not have the mathematical background to follow its development, which requires some probability theory involving counting techniques and the concept of independent events. Just this one example illustrates that the task of finding applications is quite difficult.

Once we had a focus group with some instructors who were using our books. At the end, each was to present his or her favorite application. This group was quite vocal and helpful all through the 2 days we were together, but when we got to the favorite applications, the room was total silence. No one had an application to present. This illustrates the point: Finding appealing applications that satisfy IRU is a formidable task!

I have always tried to inspire my sons by teaching them and talking to them about mathematics, in hopes that they might pursue or at least be well versed in the subject. In the case of my oldest son, Lowell, it worked because he is now an actuary with Conseco, here in Carmel, Indiana. In the case of Chris, I remember one incident while the family was riding in the car and I saw a road sign that suggested some kind of application. I said enthusiastically, "Chris, there is a mathematical formula for that application." Chris, somewhat weary of such stories responded, "Dad, you have a formula for everything!" Needless to say, one's parental attempts sometimes go awry.

## ELAINE'S PROBLEM: Does 1 = 0.99999 . . . ?

This application, simple as it may seem, is credited to my precious wife, Elaine S. Bittinger. She is a lovely, modest woman who deserves the dedication of this section to her. Living with a mathematics textbook author for almost 38 years has been a test Elaine has passed with flying colors. She labors behind the scenes giving emotional support, doing copying, running errands, but most of all being patient with my spaced-out mind. Believe it not, I rarely mow the lawn. All the men in the neighborhood are jealous! She is truly my helper-completer.

*When you look for one Bittinger, you always get the spouse as well. We're inseparable!*

Writing requires more than just sitting at a computer and putting words and symbols on paper. It often requires "writing in the mind," by which I mean the desire to find a more understandable way to present a difficult concept, yet struggling to come up with the right method. When you don't know what to do, you go through a trial-and-error process in your mind, sometimes relying on classroom experience, coauthors, reviewers, or who-knows-what to give you ideas. Then, comes that "Aha!" moment when you finally have it thought out.

I think about presenting math in the car, as I watch TV, when I'm exercising, and unfortunately when I'm supposed to be socializing with people. Many times I have gotten up at 3:30 A.M. to go to the bathroom and somehow I started thinking about a problem and had an inspiration, or even a solution. Why this happens is beyond me. Perhaps such moments are gifts from God or the workings of my subconscious. (Please don't write me letters about this, and keep your jokes to yourself.) During those "writing in the mind" situations my wife and children often catch me not listening to them when they speak. They have to bring me back. I thank them for loving me so much to accept these thousands of incidents. It was not unusual for my youngest son, Chris, to call, "Earth to Dad."

### The Problem

Inevitably, I have discussions about math with Elaine. She was an excellent, caring elementary school teacher. The problem centered, good-naturedly, around the issue of whether it is true that

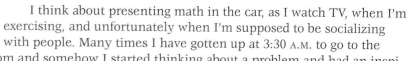

$$1 = 0.99999 \ldots . \tag{1}$$

Now the expression on the right side of this equation is actually an infinite series. But in Elaine's mind, it actually stops somewhere and becomes finite. For example, mentally to her, it might happen that

$$1 = 0.99999,$$

and this cannot be equal to 1. Now I agree with her that this statement is not true, but through 38 years of our marriage any argument I created to convince her that

$$1 = 0.99999\ldots$$

fell on deaf ears. The difficulty is the "infiniteness" of the problem. She has not been taught about the infinite. Without training, the "infinite" is a severe test of the imagination. Let's look at some of my arguments. We need some preliminary concepts.

In each of the following, the expression on the right side of the equation is a *repeating decimal*. We find repeating decimal notation from fractional notation by carrying out long division until a remainder repeats itself. For example,

$$\frac{1}{3} = 0.3333\ldots = 0.\overline{3},$$

$$\frac{2}{3} = 0.6666\ldots = 0.\overline{6}$$

$$\frac{5}{7} = 0.714285\,714285\,714285\ldots = 0.\overline{714285},$$

$$\frac{103}{17} = 6.05882352941176470588235294117647\ldots$$
$$= 6.0\overline{5882352941176470}.$$

The reader can check these divisions by hand, but it will be faster with a calculator, although it will stop with 8 to 10 digits and round. The bar on the top indicates a finite string of digits, which keeps repeating. Such a string of digits is called a *repetend*. For example, in $\frac{2}{3}$ the repetend is 6, and in $\frac{5}{7}$ it is 714285. A repetend can be thought of as if it were on a rubber stamp. Eventually, we can use the rubber stamp to keep attaching the repetend infinitely.

How long can repetends be? For $\frac{5}{97}$, the repetend has 96 digits. Note in the above equation that the repetend for $\frac{103}{17}$ has 16 digits. When we divide 103 by 17 we can have at most 16 remainders because remainders have to be less than the divisor. For any simplified fractional notation $\frac{a}{b}$, if it has a repetend, then the number of digits in the repetend is at most $b-1$. We can argue this by thinking that when we divide, we can have at most $b-1$ remainders, since they must be less than $b$. Eventually we get a repeat.

Let's look at some of the arguments I gave Elaine.

ARGUMENT 1.   Elaine, suppose you think that the decimal notation stops at some point, say after 6 digits. Then

$$1 = 0.999999,$$

but $1 \neq 0.999999$ because 0.9999999 is larger than 0.999999 and certainly closer. Elaine dismissed this argument. It simply can't be! There could have been an unwillingness at this point to allow that $0.\overline{9}$ is even symbolism for a number.

ARGUMENT 2.   Look back at the examples of repeating decimal notation for fractional notation. Elaine, do you agree that

$$\frac{1}{3} = 0.333333 \ldots \qquad = 0.\overline{3}, \tag{2}$$

and

$$\frac{2}{3} = 0.666666 \ldots \qquad = 0.\overline{6} \ ? \tag{3}$$

She totally agreed with the assertions, probably still thinking that the repetends stopped repeating. Granted these facts, what happens when we add $\frac{1}{3}$ and $\frac{2}{3}$? She quickly said we get

$$\frac{1}{3} + \frac{2}{3} = \frac{3}{3} = 1.$$

Then suppose we add the decimals on the right side:

$$\frac{1}{3} = 0.333333 \ldots \qquad = 0.\overline{3}$$

$$\frac{2}{3} = 0.666666 \ldots \qquad = 0.\overline{6}$$

$$\overline{\phantom{\frac{3}{3} = 0.999999 \ldots \qquad = 0.\overline{9} = 1}}$$

$$\frac{3}{3} = 0.999999 \ldots \qquad = 0.\overline{9} = 1$$

We should still get 1 as an answer. Do you agree? Again, she refused to accept my argument. I was dead in the water. To this day, we kid each other about this and I try these arguments again in futility.

ARGUMENT 3. At last, I found a proof that worked. It goes as follows: Elaine, do you agree that, using long division

$$\frac{1}{3} = 0.333333\ldots?$$

Remarkably, she said, "Yes." Then let's multiply on both sides by 3 and see what we get:

$$3 \cdot \frac{1}{3} = 3(0.333333\ldots)$$
$$1 \quad = 0.999999\ldots$$

Success at last! The fact that I had finally found an argument that Elaine accepted after 38 years of married life *just amazed me*. After all those intuitive "no" answers, I finally found one that worked. It was an MPX.

**MPX** MUSIC: "That's All" performed by the Purdue University Varsity Glee Club, Track 2 on the CD *Musical Memories, Vol. II*, Purdue Musical Organizations of Purdue University, www.purdue.edu/PMO/
MOVIE: *Open Range*, 2003, starring Kevin Costner, Robert Duvall, Annette Bening.
MUSIC: "I'll Never Say Goodbye" from the film *The Promise*, performed by Maureen McGovern, track 12 on the CD *The Music Never Ends*, 1997, MM Productions L.L.C., Sterling Records, Inc.

## A Proof Using Limits

ARGUMENT 4. For readers who know concepts of limits and infinite series, the following is a fourth argument. I did not impose this argument on Elaine, who did not study calculus.

The following is sigma notation for a finite sum of terms $a_k$:

$$\sum_{k=1}^{n} a_k = a_1 + a_2 + a_3 + \ldots + a_n. \tag{4}$$

The following is sigma notation for an infinite, nonterminating, sum of terms $a_k$:

$$\sum_{k=1}^{\infty} a_k = a_1 + a_2 + a_3 + \cdots \tag{5}$$

How can an infinite series converge? We let $S_n$ represent what is called a *partial sum*. That is,

$$S_n = \sum_{k=1}^{n} a_k = a_1 + a_2 + a_3 + \cdots + a_n.$$

We say that the infinite series *converges*, or has real number sum, if the limit as $n$ goes to infinity of the partial sums converges. That is,

$$\lim_{x \to \infty} S_n = S, \quad \text{where } S \text{ is a real number.}$$

For example, the following infinite series does not converge:

$$\sum_{k=1}^{\infty} 2^k = 2 + 4 + 8 + 16 + 32 + \cdots$$

But the infinite series

$$\sum_{k=1}^{\infty} \frac{1}{2^k} = \frac{1}{2} + \frac{1}{4} + \frac{1}{8} + \frac{1}{16} + \cdots$$

does converge. To see how this happens, let's consider a few partial sums:

$$S_1 = \frac{1}{2} = 0.5,$$

$$S_2 = \frac{1}{2} + \frac{1}{4} = \frac{3}{4} = 0.75,$$

$$S_3 = \frac{1}{2} + \frac{1}{4} + \frac{1}{8} = \frac{7}{8} = 0.875,$$

$$S_4 = \frac{1}{2} + \frac{1}{4} + \frac{1}{8} + \frac{1}{16} = \frac{15}{16} = 0.9375,$$

$$S_5 = \frac{1}{2} + \frac{1}{4} + \frac{1}{8} + \frac{1}{16} + \frac{1}{32} = \frac{31}{32} = 0.96875.$$

Note that when $S_n$ is expressed in fractional notation, the numerator is one less than the denominator. Since the denominator is $2^n$, it follows that

$$S_n = \frac{2^n - 1}{2^n} = 1 - \frac{1}{2^n}$$

This formula is true for all $n$. Note that as $n$ increases, the sequence

$$\frac{1}{2^n}$$

decreases rapidly toward 0. So

$$\lim_{n \to \infty} \left( 1 - \frac{1}{2^n} \right) = 1.$$

REFERENCES

Bittinger, M. L., and Morrel, B. B., *Applied Calculus,* 3rd ed, Boston, Addison-Wesley Longman, 1993, pp. 719–731.

Bittinger, M., Beecher, J., Ellenbogen, D., and Penna, J. *Algebra and Trigonometry: Graphs and Models,* 2nd ed, Boston, Addison-Wesley Longman, 2001, pp. 700–710.

Beecher, J., Penna, J., and Bittinger, M., *Algebra and Trigonometry,* Boston, Addison-Wesley Longman, 2002, pp. 697–706.

Thus,

$$S = \sum_{k=1}^{\infty} \frac{1}{2^k} = \frac{1}{2} + \frac{1}{4} + \frac{1}{8} + \frac{1}{16} + \cdots = \lim_{n\to\infty} S_n = 1$$

and the infinite series converges.

The preceding series is an example of a geometric series

$$\sum_{k=1}^{\infty} ar^k = ar + ar^2 + ar^3 + \cdots + ar^{n-1} + \cdots,$$

where $a$ is the *initial term* and $r$ is the *ratio*. From calculus it is known that a geometric series converges if

$$|r| < 1, \quad \text{that is,} \quad -1 < r < 1.$$

The sum is given by

$$\sum_{k=1}^{\infty} ar^k = ar + ar^2 + ar^3 + \cdots + ar^{n-1} + \cdots = S = \frac{a}{1-r}.$$

Now let's return to the problem in question: Does

$$1 = 0.999999\ldots?$$

On the right we have a nonterminating decimal, which is also an infinite series:

$$1 = 0.999999 \cdots$$
$$= 0.9 + 0.09 + 0.009 + \cdots.$$

This is an infinite geometric series with initial term $a = 0.9$ and ratio $r = 0.1$. Thus this series has a limit:

$$S = \frac{a}{1-r} = \frac{0.9}{1-0.1} = \frac{0.9}{0.9} = 1$$

and our proof is complete.

Now you will agree that unless Elaine takes an algebra-trig or calculus course, I cannot impose this proof upon her. But, life goes on. In truth, the infinite is very difficult to explain to anyone without enough mathematics. Experience and imagination is required to come to some understanding of the symbol ∞. It does not represent a real number.

### Pursuing Further

Instructors are welcome to copy and distribute these queries.

1. Why does the fractional notation, $\frac{a}{b}$, have to be simplified before we can assert that the number of digits in the repetend has to be less than $b$?
2. Here is another proof of 1=0.999999... See if you can complete it. Let $n=0.99999$. Then find $10n$. After that find $10n - n$, simplify both sides and solve for $n$. You will arrive at Argument 5. After accepting Argument 4, Elaine could not accept this one. It is found in many books.
3. Examine repeating decimal notation for $\frac{k}{7}$ when $k = 1, 2, 3, 4, 5,$ and 6. What do the decimals have in common?

## WENDY'S HAMBURGERS 256 WAYS

Wendy's Hamburgers ad campaigns have been interesting over the years. Do you remember "Where's the beef?" with the little old lady standing at a competitor's counter asking her question with frustration because there was not enough meat in her sandwich? Other more recent ads have featured the now deceased owner, Dave Thomas.

My favorite Wendy's sandwich is a double hamburger dripping with cheese, catsup, mustard, pickle, and lettuce. Unfortunately, that experience is an event of the past now because I had a heart attack in 1998 and abide by a very restricted low-fat diet, which is a kind of hell on earth because *I love to eat.* When I was a teenager, I could actually eat an entire pie at one sitting. When I go to heaven, the first three weeks all I want to do to is eat my favorites: hamburgers, meat loaf, pie, ice cream, pancakes, cheese omelets, and pizza. When those three weeks are over, and I will be surely satisfied, I have a long list of philosophical questions that I want to talk to God about. If nothing else, this should prove that mathematicians are not eggheads. We prioritize!

In any event, some years ago I was with my sons in a Wendy's at 55th & Keystone in Indianapolis, gorging myself with a double, when I looked up on the wall and saw an ad saying Wendy's hamburgers fixed "256 ways." I wondered, did they actually make all those hamburgers or did they use some mathematics to figure it out? As soon as I got home, I went to work on the problem.

REFERENCES
Bittinger, M., Crown, J., *Finite Mathematics,* Reading, Addison-Wesley, 1989, p. 341.
Bittinger, M., Beecher, J., Ellenbogen, D., Penna, J., *Algebra and Trigonometry: Graphs and Models,* 2nd ed, Boston, Addison-Wesley Longman, 2001, p. 767.
Beecher, J., Penna, J., and Bittinger, M., *Algebra and Trigonometry.* Boston, MA, Addison-Wesley Longman, 2002, p. 495.

Wendy's uses the following condiments for its hamburgers:

1.  catsup
2.  mustard
3.  mayonnaise
4.  tomato
5.  lettuce
6.  onions
7.  pickle
8.  cheese

Let's start with catsup. There are two ways of dealing with catsup on a hamburger, either with catsup or without. We have

$$2 = 2^1 \text{ possibilities.}$$

For each of these 2 choices, we may choose mustard or no mustard. Therefore, we have

$$2 \cdot 2 = 2^2 = 4 \text{ possibilities, and so on.}$$

Each time we add a condiment, we increase the number of possibilities by a factor of 2. We can visualize the possibilities with a number tree, as follows:

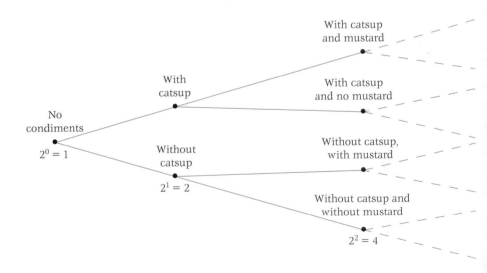

We see that if there are $n$ condiments, there are $2^n$ different possibilities for hamburgers. Since Wendy's uses 8 condiments, the total number of hamburgers is given by

$$2^8 = 256.$$

This verifies the ad.

I actually called the Wendy's home office in Columbus, Ohio, to find out how they found the result, but no one seemed to know, although I was told they actually published a list of all the possibilities. Wendy's might have missed a golden opportunity to make more of the ad campaign by including salt and pepper. Then the ad could have read

*"We fix hamburgers = $2^{10}$ = 1024 ways!"*

I think they missed the boat by not employing a mathematician. I would have taken the job for a lifetime of free hamburgers.

### Pursuing Further

1. *Pizza Hut Problem.* Pizza Hut, a national pizza firm, has the following toppings for its pizzas:
   cheese, pepperoni, sausage, mushroom, onion, green pepper, beef, Italian sausage, black olives, jalapeno peppers, ham, anchovies. How many kinds of pizza can Pizza Hut serve, excluding size and thickness of pizza?

2. *Pizza Hut Problem Expanded.* Pizza Hut serves round pizzas in sizes—9 in., 13 in., and 15 in.—and two thicknesses—thin-and-crispy and pan. Including all the toppings listed above and considering size and thickness, how many kinds of pizza can Pizza Hut serve?

3. *Arby's.* Arby's, a national firm specializing in roast beef sandwiches, actually has 26 different sandwiches ranging from roast beef to ham to turkey to chicken. Too numerous to mention, they offer 24 condiments. Including all the sandwiches and all the condiments, how many kinds of sandwich does Arby's serve?

## IT'S A SMALL, SMALL WORLD

I believe that mathematicians have some ability in all areas of the subject, but we are more talented in some areas than others. My strengths are in general topology, analysis, and logic. Although probability and statistics gave me a struggle, I have gained confidence about the theory in recent years. It has come from finding new applications of probability. Some of these are relevant and interesting, especially applications of probability in situations where the results are not what common sense might tell you.

### Mason Dots®

Made by the Tootsie Industries of Chicago, Illinois, Mason Dots is a gumdrop candy that I absolutely loved when I went to cowboy movies as a kid. Today I must avoid them for weight control. I love fruit flavors.

I opened a box of Mason Dots to begin this section and found the following number of gum drops of various colors:

REFERENCES
Bittinger, M., Crown, J., *Finite Mathematics,* Addison-Wesley Reading, MA, 1989.
Bittinger, M., Beecher, J., Ellenbogen, D., and Penna, J., *Algebra and Trigonometry: Graphs and Models,* 2nd ed, Boston, MA, Addison-Wesley Longman, 2001.
Beecher, J., Penna, J., Bittinger, M., *Algebra and Trigonometry,* Boston, MA, Addison-Wesley Longman, 2002.

| | | | |
|---|---|---|---|
| Orange | 9 | Grape | 6 |
| Lemon | 8 | Lime | 5 |
| Strawberry | 7 | Cherry | 4 |

If we put the gumdrops back in the box and mix them up thoroughly, and then reach in and without looking take one gumdrop out of the box, what is the probability of the gumdrop being lemon? lime? orange? grape? cherry? strawberry?

There are 39 gumdrops in the box. We find the probabilities by dividing the number of each flavor by 39, as the following table shows.

| Flavor | Number of That Flavor in Box | Probability |
|---|---|---|
| Orange | 9 | $\frac{9}{39} \approx 0.23$ |
| Lemon | 8 | $\frac{8}{39} \approx 0.21$ |
| Strawberry | 7 | $\frac{7}{39} \approx 0.18$ |
| Grape | 6 | $\frac{6}{39} \approx 0.15$ |
| Lime | 5 | $\frac{5}{39} \approx 0.13$ |
| Cherry | 4 | $\frac{4}{39} \approx 0.10$ |

Suppose we wanted to expand the problem to make a general prediction. That is, you walk into any candy store and buy a box of Mason Dots. What are the probabilities of getting each flavor? A statistician might conduct an experiment and examine 50 boxes and make a composite prediction. (This would be a good science fair project.) His results might then be used to predict the flavor probabilities.

If we wanted to find the probability of a particular medication's being effective, we would give the medication to a large group of people after it had been proved safe in the laboratory. Then we would keep tabs of the number of people for whom it was effective. This is *experimental* probability.

Probability theory actually began because casino owners wanted to know exactly what their chances were of winning gambling games. Mathematicians like Pascal and Fermat used reasoning to figure the probabilities. Probabilities found by reasoning are called *theoretical* probabilities.

What is true probability? In fact, both kinds of probability have validity and are commonly used. One fun aspect of probability occurs when results surprise us. Let's consider some here. I'll give them without providing proof, in the hope that they will spark an interest in your studying probability in more depth.

KISSING.   If you kiss someone who has a cold, the probability of your catching the cold is only 0.07, or 7%.

PRISON RETURN.   A person is in prison and gets released. The probability of the person committing a crime and returning to prison is 0.80, or 80%.

Each of the preceding probabilities was calculated as experimental probability. The following was proved by one of my coauthors, J. Conrad Crown.

SMALL WORLD PROBLEM.   Have you ever been away from home, met a stranger, started talking about people you know, discovered a common acquaintance, and exclaimed, "It's a small world!" Actually, it is not as small as you think. The probability of this happening is 0.20, or 20%, much more than the virtual 0 we might have guessed.

THE BIRTHDAY PROBLEM REVISITED.   Of $n$ people in a group, what is the probability that at least 2 of them have the same birthday (day and month, but not necessarily the same year)?

We considered this problem earlier, and now we're going to give more detail. Let $E$ = the event of 2 people in a group of $n$ people having the same birthday. The probability of $E$ occurring is given by

$$p(E) = 1 - \frac{365}{365} \cdot \frac{364}{365} \cdot \frac{363}{365} \cdots \frac{[365 - (n-1)]}{365}$$

The following table and graph show results for groups of size 2 to 100. For 23 or more people, the probability that 2 or more people will have the same birthday is greater than $\frac{1}{2}$, that is, it is more likely to occur than not!

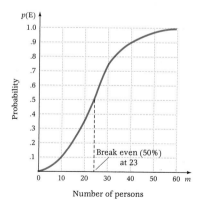

| $n$ | $p(E)$ | $n$ | $p(E)$ | $n$ | $p(E)$ | $n$ | $p(E)$ | $n$ | $p(E)$ |
|---|---|---|---|---|---|---|---|---|---|
| 2 | 0.00274 | 16 | 0.284 | 24 | 0.538 | 32 | 0.753 | 40 | 0.891 |
| 5 | 0.0271 | 17 | 0.315 | 25 | 0.569 | 33 | 0.775 | 50 | 0.970 |
| 10 | 0.117 | 18 | 0.347 | 26 | 0.598 | 34 | 0.795 | 60 | 0.9941 |
| 11 | 0.141 | 19 | 0.379 | 27 | 0.627 | 35 | 0.814 | 70 | 0.99916 |
| 12 | 0.167 | 20 | 0.411 | 28 | 0.654 | 36 | 0.832 | 80 | 0.999914 |
| 13 | 0.194 | 21 | 0.444 | 29 | 0.681 | 37 | 0.849 | 90 | 0.999994 |
| 14 | 0.223 | 22 | 0.476 | 30 | 0.706 | 38 | 0.864 | 100 | 0.9999997 |
| 15 | 0.253 | 23 | 0.507 | 31 | 0.730 | 39 | 0.878 | | |

### Pursuing Further

*Linguistics.* An experiment was conducted by the author to determine the relative occurrence of various letters of the English alphabet. The front page of a newspaper was considered. In all, there were 9,136 letters. The number of occurrences of each letter of the alphabet is listed in the following table.

1. Complete the table of probabilities with the percentage, to the nearest tenth of a percent, of the occurrence of each letter.
2. What is the probability of a vowel occurring?
3. What is the probability of a consonant occurring?

| Letter | Number of Occurrences | Probability |
|--------|-----------------------|-------------|
| A | 853 | 853/9136 ≈ 9.3% |
| B | 136 | |
| C | 273 | |
| D | 286 | |
| E | 1229 | |
| F | 173 | |
| G | 190 | |
| H | 399 | |
| I | 539 | |
| J | 21 | |
| K | 57 | |
| L | 417 | |
| M | 231 | |
| N | 597 | |
| O | 705 | |
| P | 238 | |
| Q | 4 | |
| R | 609 | |
| S | 745 | |
| T | 789 | |
| U | 240 | |
| V | 113 | |
| W | 127 | |
| X | 20 | |
| Y | 124 | |
| Z | 21 | 21/9136 ≈ 0.2% |

4. Compare the QWERTY and DVORAK typewriter/computer keyboards. How can you use a linguistics table to show why DVORAK created his keyboard?

5. *Wheel of Fortune®.* The results of this alphabet counting experiment can be quite useful to a person playing the popular television game show *Wheel of Fortune.* Players guess letters in order to spell a phrase, a person, or a thing.
   a. What 5 consonants have the greatest probability of occurring?
   b. What vowel has the greatest probability of occurring?
   c. The winner of the main part of the show plays for a grand prize and at one time was allowed to guess 5 consonants and a vowel in order to discover the secret wording. The 5 consonants *R, S, T, L, N,* and the vowel *E* seemed to be chosen most often. Do the results in parts (a) and (b) support such a choice?

6. *Sociological Survey.* Your author's son, Lowell, conducted a science fair experiment to determine the number of people who are left-handed, right-handed, or both. The results are shown in the graph at left.
   a. Determine the probability that a person is right-handed.
   b. Determine the probability that a person is left-handed.
   c. Determine the probability that a person is ambidextrous (uses both hands with equal ability).
   d. For most tournaments held by the Professional Bowlers Association, there are 120 bowlers. On the basis of the data in this experiment, how many of the bowlers would you expect to be left-handed?

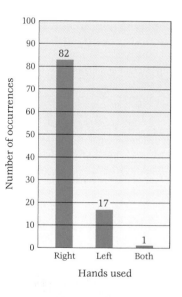

7. *Quality Control.* It is very important for a manufacturer to maintain the quality of its products. In fact, companies hire quality control inspectors to ensure this process. The goal is to produce as few defective products as possible. But since a company is producing thousands of products every day, it cannot afford to check every product to see if it is defective. To find out what percentage of its products are defective, the company checks a smaller sample.

    The U.S. Department of Agriculture requires that 80% of the seeds that a company produces must sprout. To find out about the quality of the seeds it produces, a company takes 500 seeds from those it has produced and plants them. It finds that 417 of the seeds sprout.
   a. What is the probability that a seed will sprout?
   b. Did the seeds pass government standards?

8. *Television Ratings.* Television networks are always concerned about the percentage of homes that have TVs and are watching their programs. A sample of the homes are contacted by attaching an electronic device to the TVs of about 1,400 homes across the country. Viewing information is then fed into a computer. The following table shows the results of a recent survey.

| Network | ABC | CBS | NBC | Fox | Other, or Not Watching |
|---|---|---|---|---|---|
| Number of Homes Watching | 118 | 134 | 143 | 99 | 906 |

What is the probability that a home was tuned to NBC during the time period? to Fox?

## UP, UP, AND AWAY!

Phrases in ordinary conversation often refer to exponential functions. For example, "The stock market rose at an exponential rate" or "Rabbits breed exponentially."

Graphs in newspapers and magazines often display exponential functions, such as the following:

| Super Bowl | Year | Price of a Ticket to the Super Bowl |
|---|---|---|
| I | 1967 | $12 |
| XI | 1977 | $20 |
| XXI | 1987 | $75 |
| XXXI | 1997 | $275 |
| XXXV | 2001 | $325 |
| XXXVI | 2002 | $400 |

*Source:* National Football League

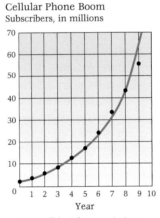

Cellular Phone Boom
Subscribers, in millions

*Source:* Cellular Telecommunications Industry Association

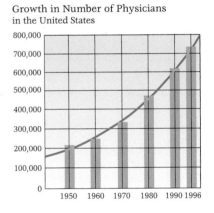

Growth in Number of Physicians in the United States

*Source:* American Medical Association

Such statements and graphs refer to a rapidly rising situation that goes on without end. If you own stocks you would like to see their value increase exponentially. But farmers would like some controls on rabbit population growth. Very often exponential increases are uncontrollable.

The theory of exponential functions $f(x) = Ba^{kx}$, $a > 0$ has always fascinated me because it holds together in such a beautiful structure and is so rich in applications. I enjoy this theory so much that I even named my softball team the "Exponents." My shirt number is 8, because $8 = 2^3$ and because there are so many geometric symmetries in the numeral.

Many real-world examples of exponential functions are found in textbooks on college algebra and in calculus (see the references). We will consider a few briefly later, but for now let's look at a fun example from the movies that even has moral implications.

The movie "Pay It Forward," starring Kevin Spacey, Helen Hunt, and Haley Joel Osment (based on the novel of that name by C. R. Hyde), suggests an application of exponential functions. Osment's character, Trevor McKinney, does an act of kindness to each of three people. In return, each of these people must agree to "pay forward" three acts of kindness to other people, and so on. Let's assume that each person performs these acts of kindness within 1 month.

We can find an exponential function for the number, $N$, of people who have received the acts of kindness, in terms of $t$, where $t =$ time in months. We can think of the process using a number tree very much like the way we did the Wendy's hamburger application. In fact, Trevor presents the problem in that way to his classmates and his teacher, Eugene Simonet, played by Kevin Spacey.

Let's start with Trevor's 3 acts of kindness. We then have

$$3 = 3^1 \text{ acts of kindness.}$$

After 1 month each of these people does 3 more acts of kindness. We then have

$$9 = 3^2 \text{ acts of kindness.}$$

After another month each of these 9 people does an act of kindness, resulting in

$$27 = 3^3 \text{ acts of kindness, and so on.}$$

Compiling the results in the following table, we get

| Month, $t$ | Acts of Kindness |
|:---:|:---:|
| 1 | 3 |
| 2 | 9 |
| 3 | 27 |
| 4 | 81 |
| 5 | 324 |
| 12 | 531,441 |
| $t$ | $3^t$ |

Then we look for a pattern.

We see that after $t$ months, $t \geq 1$, another $3^t$ acts of kindness are performed. The exponential function

$$K(t) = 3^t, \ t > 1,$$

describes the application.

REFERENCES

Bittinger, M., Beecher, J., Ellenbogen, D., and Penna, J., *Algebra and Trigonometry: Graphs and Models,* 2nd ed, Boston, Addison-Wesley Longman, 2001, pp. 282–358.

Beecher, J., Penna, J., and Bittinger. M., *Algebra and Trigonometry,* Boston, Addison-Wesley Longman, 2002, pp. 262–334.

Bittinger. M. L., *Intermediate Algebra,* 9th ed, Boston, Addison-Wesley Longman, 2003, pp. 686–775.

Bittinger, M. L., *Calculus and Its Applications,* 8th ed, 2004, pp. 285–362.

Thus, after the 12th month, there are $3^{12}$, or 531,441, acts of kindness performed. If we add all the acts together, we get 797,160. The population of the world is about 6.2 billion. Since,

$$3^{20} = 3{,}486{,}784{,}401, \quad \text{and}$$
$$3^{21} = 10{,}460{,}353{,}203,$$

we see that even if we ignored the cumulative number, the entire world would receive acts of kindness within 2 years, assuming no repeats.

If you have seen the movie, you know that Trevor dies at the end. We then see an endless stream of carlights moving toward his funeral, testifying to the exponential effect of Trevor's 3 initial acts of kindness. This is certainly consistent with the exponential growth of Trevor's function.

In the remainder of this section, we will briefly consider some of the theory and applications of exponential functions first from the standpoint of algebra and trigonometry, and then from calculus.

### Exponential Growth via Algebra and Trigonometry

POPULATION GROWTH.    The function

$$P(t) = P_0 e^{kt}, \ k > 0, \ e \approx 2.7182818 \ldots,$$

is a model of many kinds of population growth, whether it be a population of acts of kindness, people, bacteria, cellular phones, or money. In this function $P_0$ is the population at time, $t = 0$, $P(t)$ = the population after time $t$, and $k$ is called the *exponential growth rate.* The graph of such an equation is as follows.

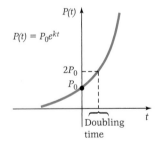 

Here are some examples of real-life applications of this type of exponential growth.

POPULATION GROWTH OF THE UNITED STATES. In 2000 the population of the United States was about 281 million, and the exponential growth rate was 0.8% per year.

The exponential growth function is $P(t) = 281e^{0.008t}$, where $t$ = the number of years after 2000.

The population in the year 2012 can be predicted from the function:
$P(12) = 281e^{0.008(12)} \approx 309$ million.

The time for a population or an investment, growing exponentially, to double itself is called the *doubling time*. The doubling time is given by $T = \frac{\ln 2}{0.008} = 86.6$ years. We find this formula by solving the equation $562 = 281e^{0.008T}$ for $T$. It turns out that the doubling time depends solely on the exponential growth rate, not the initial value, of the population. Thus, 281 million grows to 562 million in 86.6 years, 562 million grows to 1,124 million in 86.6 years, and so on.

The connection between doubling time and an exponential growth rate is given as follows.

---

### Growth Rate and Doubling Time

The **growth rate** $k$ and the **doubling time** $T$ are related by

$$kT = \ln 2, \quad \text{or} \quad k = \frac{\ln 2}{T}, \quad \text{or} \quad T = \frac{\ln 2}{k}.$$

---

For example, we know that the population of the world is doubling about every 54.6 years. We can find the exponential growth rate as follows:

$$k = \frac{\ln 2}{T} = \frac{\ln 2}{54.6} \approx 1.3\%.$$

Thus, the growth rate of world population is about 1.3% per year.

The rather simple result relating doubling time to growth rate just amazed me when I first learned it. If you know the growth rate, you can find the doubling time by dividing it into ln 2 using the formula $=\frac{\ln 2}{k}$. If you know the doubling time, you can find the growth rate by dividing it into ln 2 again.

I was so impressed and amazed by this connection between doubling time and interest rate that I call it a Math Power Experience.

**MPX**   MUSIC: Shostakovich's 5th Symphony in D Minor, Opus 47, IV-Allegro non Troppo, recorded on CD by Leonard Bernstein and the New York Philharmonic Orchestra.

INTEREST COMPOUNDED CONTINUOUSLY. The exponential growth function $P(t) = P_0e^{kt}$ is also a model for the growth of money invested at interest rate k, where interest is *compounded continuously*. For example, at 8% interest compounded continuously, the amount of $2000 will grow after $t$ years to an amount given by $P(t)=\$2000e^{0.08t}$. The balance in the investment after 10 years is

$$P(10)=2000e^{0.08(10)} \approx \$4451.08.$$

The doubling time of the money is given by

$$T = \frac{\ln 2}{k} = \frac{\ln 2}{0.08} \approx 8.7 \,\text{yr.}$$

THE RULE OF 69.   The relationship between doubling time $T$ and interest rate $k$ is the basis of a rule often used in the investment world called the Rule of 69. To estimate how long it will take to double your money at varying rates of return, divide 69 by the rate of return. To see how this works, let the interest rate $k = r\%$. Then

$$T = \frac{\ln 2}{k} \approx \frac{0.693147}{r\%} = \frac{0.693147}{r \times 0.01} \cdot \frac{100}{100} = \frac{69.3147}{r} \approx \frac{69}{r}.$$

Suppose the interest rate is 6.2%, then the doubling time is about

$$\frac{69}{r} = \frac{69}{6.2} \approx 11.1 \,\text{yr.}$$

THE VALUE OF A 1929 PIERCE-ARROW ROADSTER.   My uncle, Dr. Rex B. Gosnell, my mother's brother, was my educational hero. He got his B.S. from Manchester College and went to Purdue University for his Ph.D. in chemistry. I did the same in mathematics education, with a stop at Ohio State for an M.S. in mathematics. Over the years, when I would visit him in San Diego, I would see various stages of restoration of a very popular classic car, a 1929 Pierce-Arrow Roadster, shown in the picture. Rex and his wife Alice recently donated his car to the Classic Car Club of America Museum in

*Dr. and Mrs. Rex B. Gosnell (the uncle and aunt of your author) donated this 1929 Pierce-Arrow Roadster to the Classic Car Club of America Museum in Hickory Corners, Michigan.*

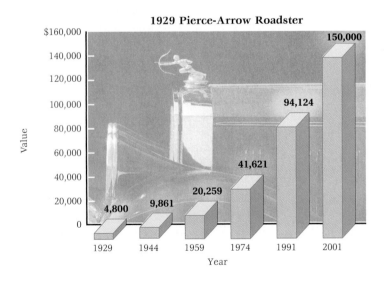

Hickory Corners, Michigan, near Grand Rapids. It occurred to me at the time of this donation that an exponential function might be used to model the value of the car over the years.

*The value of a 1929 Pierce-Arrow Roadster was $4,800 when it was new. Fully restored, the value of this automobile in 2001 was $150,000.*

*The value of this car t years after it was restored can be modeled by the exponential function* $P(t) = \$4800e^{0.048t}$. *It can then be shown that its value will be $202,880 in the year 2007, and that the doubling time of its value is about 14.4 years.*

### Pursuing Further

1. *Price of a Super Bowl Ticket.* Every year around Super Bowl time there are discussions of the price of a ticket, together with tables of past prices. When you make a graph of the data it is fun to see the exponential growth. The price of a ticket to the first Super Bowl in 1967 was $12. By 2002 it rose to $400.
   **a.** If we let $t = 0$ correspond to 1967 and $t = 35$ correspond to 2002, then $t =$ the number of years since 1967. Use the data points (0, 12) and (35, 400) to find the exponential growth rate and fit an exponential growth function to the data.
   **b.** Use the function found in part (a) to predict the price of a ticket to the Super Bowl in 2010 and 2020.
   **c.** Assuming exponential growth continues, predict the year in which the price of a ticket will be $1,000, then $2,000.
2. *Find Your Own Application.* Almost every day you can find data in the newspaper that fits exponential growth. Such situations reinforce "what math is good for," such as using exponential functions.
   Check newspapers and magazines looking for data with fast-rising growth that seems to fit an exponential function. How hard was it to find the data? Use your graphing calculator to fit an exponential function to the data. Then make predictions. What is the doubling time?
3. Over the past few years there have been three movies featuring stories of brilliant math students or professors. Can you name them?
4. There is an excellent play, *Proof,* by David Auburn that had a successful run on Broadway and is now on national tour throughout the country. Try to see it.
5. Believe it or not, A. G. Reinhold has a Web site devoted to "The Math in the Movies Page: A Guide to Major Motion Pictures with Scenes of Real Mathematics." The URL is http://world.std.com/~reinhold/mathmovies.html
   Check it out, pick a movie, and try to see it.

REFERENCE
Bittinger, M. L., *Calculus and Its
   Applications,* 8th ed, Boston,
   Addison-Wesley Longman,
   2004.

# THE DEAD CALCULUS PROFESSOR

I first discovered this problem in a professional journal. It intrigued me because it involved exponential functions and was relevant to what a coroner might do to determine the time of death in a murder case. Indeed, I called a funeral director to discuss some of the numbers in the problem. It involves Newton's Law of Cooling.

WHEN WAS THE MURDER COMMITTED?   The police discover the body of a calculus professor. Critical to solving the crime is determining when the murder was committed. The police call the coroner, who arrives at 12 noon. She immediately takes the temperature of the body and finds it to be 94.6°. She waits 1 hour, takes the temperature again, and finds it to be 93.4°. She also notes that the temperature of the room is 70°. When was the murder committed?

A hot cup of soup, at a temperature of 200°, is placed in a room whose temperature is 70°. The temperature of the soup cools over time $t$, in minutes, according to the mathematical model, or equation, called *Newton's Law of Cooling.*

The temperature $T$ of a cooling object drops at a rate that is proportional to the difference $T - C$, where $C$ is the constant temperature of the surrounding medium. Thus,

$$\frac{dT}{dt} = -k(t - C). \qquad (6)$$

The function that satisfies equation (6) is

$$T = T(t) = ae^{-kt} + C, \text{ where } t \text{ is time.} \qquad (7)$$

The graph of $T(t) = ae+^{-kt} + C$ shows that $\lim_{t \to \infty} T(t) = C$. The temperature of the object decreases toward the temperature of the surrounding medium.

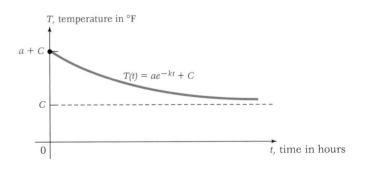

Mathematically, this model tells us that the temperature never reaches $C$, but in practice this happens eventually. At least, the temperature of the cooling object gets so close to that of the surrounding medium that no device could detect a difference. Let's now see how Newton's Law of Cooling can be used to solve the murder.

We first find $a$ in the equation $T(t) = ae^{-kt} + C$. Assuming that the temperature of the body was normal when the murder occurred, we have $T = 98.6°F$ at $t = 0$. Thus,

$$98.6 = ae^{-k\cdot 0} + 70, \text{ so}$$

$$a = 28.6.$$

Thus, $T$ is given by $T(t) = 28.6e^{-kt} + 70$.

We want to find the number of hours $N$ since the murder was committed. To do so, we must first determine $k$. From the two temperature readings the coroner made, we have

$$94.6 = 28.6e^{-kN} + 70, \quad \text{or} \quad 24.06 = 28.6e^{-kN}; \tag{8}$$

$$93.4 = 28.6e^{-k(N+1)} + 70, \quad \text{or} \quad 23.4 = 28.6e^{-k(N+1)}. \tag{9}$$

Dividing equation (8) by equation (9), we get

$$\frac{24.6}{23.4} = \frac{28.6e^{-kN}}{28.6e^{-k(N+1)}}$$

$$= e^{-kN+k(N+1)}$$

$$= e^{-kN+kN+k} = e^k.$$

We solve this equation for $k$:

$$\ln \frac{24.6}{23.4} = \ln e^k \qquad \text{Taking the natural logarithm on both}$$

$$0.05 \approx k. \qquad \text{sides}$$

Next, we substitute back into equation (8) and solve for $N$:

$$24.6 = 28.6e^{-0.05N}$$

$$\frac{24.6}{28.6} = e^{-0.05N}$$

$$\ln \frac{24.6}{28.6} = \ln e^{-0.05N}$$

$$-0.150660 \approx -0.05N$$

$$3 \approx N.$$

Since the coroner arrived at 12 noon, the murder was committed at about 9:00 A.M.

When I talked to the funeral director about the numbers in this problem and the feasibility of using Newton's Law of Cooling, he explained that in truth the temperature of a dead body actually rises above the normal of 98.6°, but for such a very short time that the cooling formula still can be used as a model. The graph might look like the following:

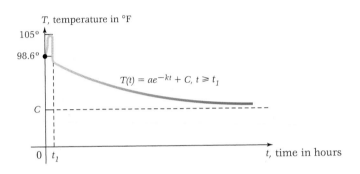

### Pursuing Further

1. *Cooling Body.* The coroner arrives at the scene of a murder at 11 P.M. She takes the temperature of the body and finds it to be 85.9°. She waits 1 hour, takes the temperature again, and finds it to be 83.4°. She notes that the room temperature is 60°. When was the murder committed?

2. *Newton's Law of Cooling.* Consider the following exploratory situation. Draw a glass of hot tap water. Place a thermometer in the glass and check the temperature. Check the temperature every 30 minutes thereafter. Plot your data on this graph, and connect the points with a smooth curve.

   **a.** What was the temperature when you began?

   **b.** At what temperature does there seem to be a leveling off of the graph?

   **c.** What is the difference between your answers to parts (a) and (b)?

   **d.** How does the temperature in part (b) compare with the room temperature?

   **e.** Find an equation that "fits" the data. Use this equation to check values of other data points. How do they compare?

   **f.** Is it ever theoretically possible for the temperature of the water to be the same as the room temperature? Explain.

   **g.** Find the rate of change of the temperature and interpret its meaning.

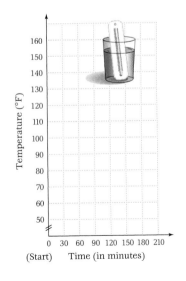

## THE TOWER OF HANOI PROBLEM

In the preceding section we enjoyed finding applications of math to real-world problems. In this section we consider an application of math that is enjoyable from a purely mathematical standpoint. Students need such experiences if they are to develop a passion for math.

One of my favorite methods of proof is mathematical induction, (MI), a topic usually introduced to students near the end of an algebra-trig course, although it may often be omitted because of time constraints. I like MI because the second stage of such proofs usually necessitates an elegant moment of insight. Although the logic of such proofs can sometimes be difficult to get across to students the first time they encounter it, MI is a very important proof tool in advanced mathematics. We will explain and exemplify one such proof.

The Tower of Hanoi Puzzle was originally developed in 1883 by E. Lucas, a professor at Lycée Saint-Louis in France. There are three pegs on a board. On one peg are *n* disks, each smaller than the one on which it rests. The puzzle is to move this pile of disks to another peg. The final order must be the same, but you can move only one disk at a time and can never place a larger disk on a smaller one.

We proceed with this problem in three stages. To build our understanding of the problem, let's first determine the *smallest* number of moves needed to move 1 disk, 2 disks, 3 disks, and 4 disks. Then we will step back and look for a pattern.

Suppose we move only 1 disk. We label the pegs *A*, *B*, and *C*, and assume the disk starts on *A*. Then the smallest number of moves of 1 disk to another is simply 1. See Figure 1. We take 1 disk off the pile and move it to another peg. (We would get more moves if we simply moved it continuously, but we want the fewest moves.)

FIGURE 1

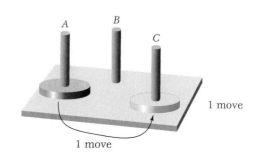

REFERENCES

Bittinger, M., Beecher, J., Ellenbogen, D., and Penna, J., *Algebra and Trigonometry: Graphs and Models,* 2nd ed, Boston, Addison-Wesley, 2001, p. 767ff.

Bittinger, M., *Logic, Proof, and Sets,* Addison-Wesley, Reading, MA, 1982, p. 63. This book is no longer in print.

Suppose we are to move 2 disks. To ease notation, we think of the smallest disk as $x$ and the next larger as $y$. We move $x$ to $B$, $y$ to $C$, and $x$ on top of $y$ on $C$. The smallest number of moves is 3, as Figure 2 shows.

FIGURE 2

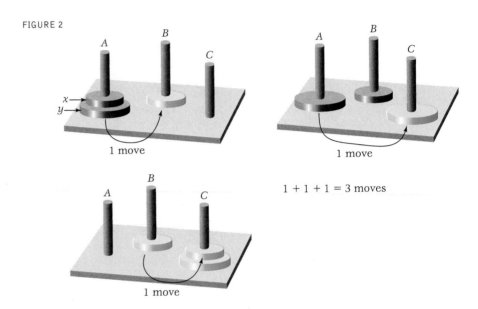

1 move

1 move

1 + 1 + 1 = 3 moves

1 move

Suppose we are to move 3 disks. Now the task gets more formidable. The moves are described in Figure 3.

FIGURE 3

**Move:**
**1.** $x$ onto $B$

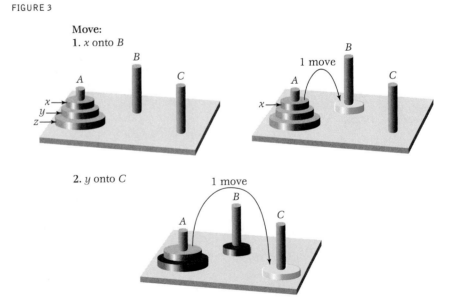

1 move

**2.** $y$ onto $C$

1 move

**3.** $x$ from $B$ on top of $y$ on $C$

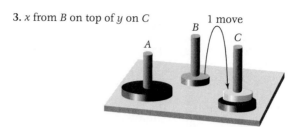

**4.** $z$ on $B$

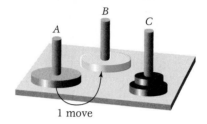

**5.** $x$ back to $A$ from $C$

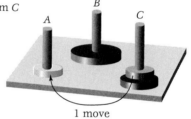

**6.** $y$ on top of $z$ on $B$

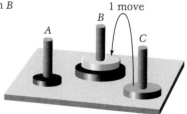

**7.** $x$ from $A$ on top of $y$ on top of $z$ on $B$

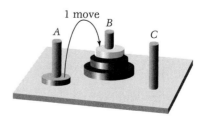

REFERENCE

Polya, G. *How to Solve It: A New Aspect of Mathematical Method,* Princeton University Press; Princeton, NJ, softcover reissue (Nov. 1, 1971), hardcover reissue (Jan. 1982).

Have you been doing these on your own? Try doing 4 disks on your own. See if you can discover a quick way to think out the task.

In his famous book *How To Solve It,* George Polya suggests an idea for problem solving that bears much fruit. When faced with a tough task, first try a smaller, less formidable case. We will do that twice with this problem, once here as we try to calculate the moves for 4 disks, and again in the proof by mathematical induction.

Let's organize the results for 1 through 3 disks in a table and look for a pattern. Have you discovered and conjectured a formula yet? Let's assume you have not, but you don't want to carry out all the moves one at a time to move 4 disks. Notice that in the pile of 4 disks, we have 3 disks on top. Look at Figure 4.

FIGURE 4

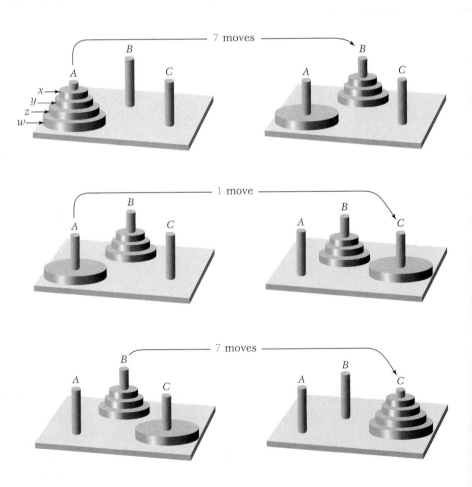

Ignoring the bottom disk, we know that it will take 7 moves to get the 3-disk pile to another disk. So let's put all 3 over on peg $B$. Then we take the bottom disk $w$ off $A$ and move it to peg $C$. That's 1 move. Finally, we move the 3 disks off peg $B$ on top of $w$ on peg $C$. That takes 7 moves. The total number of moves is then

$$7 + 1 + 7 = 15.$$

We can then add this fact to our chart.

| Number of Disks | Fewest Moves |
|:---:|:---:|
| 1 | 1 |
| 2 | $1 + 1 + 1 = 3 = 4 - 1 = 2^2 - 1$ |
| 3 | $3 + 1 + 3 = 7 = 8 - 1 = 2^3 - 1$ |
| 4 | $7 + 1 + 7 = 15 = 16 - 1 = 2^4 - 1$ |
| $\vdots$ | $\vdots$ |
| $n$ | $(2^{n-1} - 1) + 1 + (2^{n-1} - 1) = 2^n - 1$ |

Have you conjectured a formula for the smallest number of moves of $n$ disks? Perhaps you have discovered the following formula:

---

### The Tower of Hanoi Formula

The *smallest* number of moves needed to move $n$ disks is

$$2^n - 1.$$

---

Our third task is to prove the conjecture using mathematical induction (MI). MI is used to prove that all statements in an infinite sequence of statements are true. The statements usually have the form:

For all natural numbers $n$, $S_n$,

where $S_n$ is some mathematical sentence such as those of the preceding examples. Of course, we cannot prove each statement of an infinite sequence individually. Instead, we try to show that "whatever $S_k$ holds, then $S_{k+1}$ must hold." We abbreviate this as $S_{k+1} \rightarrow S_{k+1}$. (This is also read "If $S_k$, then $S_{k+1}$" or "$S_k$ implies $S_{k+1}$.") Suppose that we could somehow establish that this holds for all natural numbers $k$. Then we would have the following:

$S_1 \rightarrow S_2$ meaning "if $S_1$ is true, then $S_2$ is true";

$S_2 \rightarrow S_3$ meaning "if $S_2$ is true, then $S_3$ is true";

$S_3 \rightarrow S_4$ meaning "if $S_3$ is true, then $S_4$ is true"; and so on, indefinitely.

Even knowing that $S_k \to S_{k+1}$, we would still not be certain whether there is *any* $k$ for which $S_k$ is true. All we would know is that "if $S_k$ is true, then $S_{k+1}$ is true." Suppose now that $S_k$ is true for some $k$, say $k = 1$. We then must have the following.

$S_1$ is true.    We have verified, or proved, this.

$S_1 \to S_2$    This means that whenever $S_1$ is true, $S_2$ is true.

Therefore, $S_2$ is true.

$S_2 \to S_3$    This means that whenever $S_2$ is true, $S_3$ is true.

Therefore, $S_3$ is true,

and so on.

We conclude that $S_n$ is true for all natural numbers $n$.

This leads us to the principle of mathematical induction, which we use to prove the types of statement considered here.

---

### The Principle of Mathematical Induction

We can prove an infinite sequence of statements $S_n$ by showing the following.

1.  *Basis step.* $S_1$ is true.
2.  *Induction step.* For all natural numbers $k$, $S_k \to S_{k+1}$.

---

Mathematical induction is analogous to lining up a sequence of dominoes. The induction step tells us that if any domino is knocked over, then the one next to it will be hit and knocked over. The basis step tells us that the first domino can indeed be knocked over. Note that in order for all dominoes to fall, *both* conditions must be satisfied.

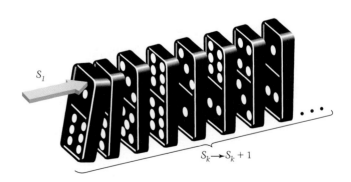

When doing proofs by mathematical induction, it is helpful to first write out $S_n$, $S_1$, $S_k$, and $S_{k+1}$. This helps to identify what is to be assumed. Let $P_n =$ the least number of moves required to move $n$ disks. We must show:

$$S_n : P_n = 2^n - 1.$$

Then we take the *basis step*: $S_1 = 2^1 = 2 - 1 = 1$, followed by the *induction step*: Assume $S_k$ for $k$ disks:

$$P_k : 2^k - 1.$$

Then prove $S_{k+1}$ for $k+1$ disks:

$$P_{k+1} = 2^{k+1} - 1.$$

The remainder of the proof involves essentially the same thinking we did to go from 3 disks to 4 disks. Suppose there are $k + 1$ disks on a peg. We know from $S_k$ that we can move $k$ of them to another disk in $2^k - 1$ moves. Then we move the remaining large disk to another peg in 1 move. Then we can move the other $k$ disks back on it in another $2^k - 1$ moves. Thus, the total number of moves $P_{k+1}$ is

$$(2^k - 1) + 1 + (2^k - 1) = 2\,(2^k - 1) + 1 = 2 \cdot 2^k - 2 + 1 = 2^{k+1} - 1.$$

We have deduced $S_{k+1}$ from $S_k$ and our proof is complete.

The beauty of this proof to me is how we can discover a "quick" way of counting the moves, as in the case of 4 disks, and then extrapolate that to the proof in the induction step. Making such a connection was indeed an **MPX**.

**MPX** MUSIC: "How the West Was Won" by Alfred Newman, track 6 on the CD, *Round-Up*, recorded by Erich Kunzel and the Cincinnati Pops Orchestra, © 1986 TELARC Digital.

### *Pursuing Further*

Prove each by mathematical induction.

1. For every natural number $n$,

$$1 + 3 + 5 + \cdots + (2n - 1) = n^2.$$

2. For every natural number $n$,

$$1^2 + 2^2 + 3^2 + \cdots + n^2 = \frac{n(n + 1)(2n + 1)}{6}.$$

3. For every natural number $n$,

$$\frac{1}{2} + \frac{1}{4} + \frac{1}{8} + \cdots + \frac{1}{2^n} = \frac{2^n - 1}{2^n}.$$

**4.** For every natural number $n$, 2 is a factor of $n^2 + n$.

**5.** For every natural number $n$, $|\sin(nx)| \leq |n||\sin x|$.

## MY GOLF PROBLEMS

I have made many feeble attempts at becoming a good golfer. I tell people that God has given me many blessings in life, but golf is not one of them because I can almost hear a "still, small voice" saying, "I am not giving this to you, Marv! You will spend too much time on golf and lose touch with your family and your real talents!" Nevertheless, I keep trying. I do enjoy learning about the game, being out in nature, and the solitude. I never take my telephone. But even on the golf course I find mathematics.

*Golf balls in a pile with a square base. (Photo credit: Karen E. Bittinger)*

One sometimes encounters stacks of balls arranged in layers that form a pyramidal shape like that of a stack of cannon balls, a stack of oranges in a supermarket, and a stack of golf balls on a driving range. My coauthor, Judy Beecher, and I were intrigued to discover a polynomial function that can be used to predict quickly the exact number of balls in the pile.

We soon discovered that we could conceive the problem, but physical stacks of balls like these fall apart from their own weight. One needs a ridge along a container at the bottom of the balls. Look carefully at the golf ball tray and you will see a ridge at the bottom of the pile.

I discovered these trays when I went to a driving range on one of my favorite courses, Golf Club at Eagle Mountain in Fountain Hills, Arizona. That started a marathon of phone calls until I finally found the trays. The company sold me the trays but could not understand why I didn't want to buy a machine to fill them.

We start to solve this problem by giving a formula and proving it later.

### Stacks of Golf Balls

A stack of golf balls is formed by successive layers of squares of golf balls. The number $N$ of balls in the stack is given by the polynomial function

$$N(x) = \frac{1}{3}x^3 + \frac{1}{2}x^2 + \frac{1}{6}x,$$

where $x$ = the number of layers. The graph of the function is shown below.

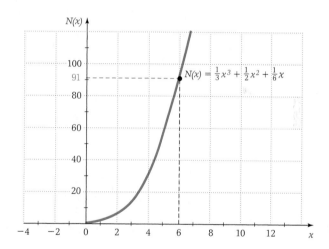

For a stack with 6 layers, there is a total of 91 golf balls, as seen from the graph and by finding the function value $N(6)$.

$$N(x) = \frac{1}{3}x^3 + \frac{1}{2}x^2 + \frac{1}{6}x,$$
$$N(6) = \frac{1}{3} \cdot 6^3 + \frac{1}{2} \cdot 6^2 + \frac{1}{6} \cdot 6$$
$$= 72 + 18 + 1$$
$$= 91.$$

### Proof of the Formula

Let's prove the formula. We look at each successive layer from the top down. Look for a pattern in the following:

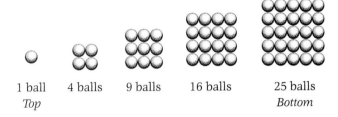

| 1 ball | 4 balls | 9 balls | 16 balls | 25 balls |
| *Top* | | | | *Bottom* |

The sum of all the balls in the 5 stacks is $1 + 4 + 9 + 16 + 25$, a sum of squares:

$$1^2+2^2+3^2+4^2+5^2.$$

What if we had $x$ layers? We might conjecture that if there are $x$ layers, the sum of all the layers is

$$1^2+2^2+3^2+ \cdots + x^2.$$

We know another formula from our introduction. Therefore, we hypothesize that if there are $x$ layers, then the total number of golf balls is

$$1^2 + 2^2 + 3^2 \cdots + x^2 = \frac{1}{3}x^3 + \frac{1}{2}x^2 + \frac{1}{6}x$$
$$= \frac{x(x+1)(2x+1)}{6}$$

How can we prove this conjecture? Using $n$ for $x$ we can rewrite it as

$$1^2 + 2^2 + 3^2 + \cdots + n^2 = \frac{n(n+1)(2n+1)}{6},$$

which is true for all natural numbers. The formula holds as well for all real numbers. We prove it here for natural numbers by mathematical induction.

### Proof by Mathematical Induction

Prove that for all natural numbers $n$, $\quad \sum_{j=1}^{n} j^2 = \frac{n(n+1)(2n+1)}{6}.$

We must show $\quad S_n : \sum_{j=1}^{n} j^2 = \frac{n(n+1)(2n+1)}{6}.$

1. *Basis step:* Prove $\quad S_1 : \sum_{j=1}^{1} j^2 = \frac{1(1+1)(2 \cdot 1+1)}{6}.$

This follows by doing the calculations on each side.

2. *Induction step*

$$\text{Assume} \quad S_k : \sum_{j=1}^{k} j^2 = \frac{k(k+1)(2k+1)}{6}.$$

$$\text{Deduce} \quad S_{k+1} : \sum_{j=1}^{k+1} j^2 = \frac{(k+1)[(k+1)+1][2(k+1)+1]}{6}, \quad \text{or,}$$

simplifying the right side we get

$$S_{k+1} : \sum_{j=1}^{k+1} j^2 = \frac{(k+1)(k+2)(2k+3)}{6}.$$

Now

$$\sum_{j=1}^{k+1} j^2 = \sum_{j=1}^{k} j^2 + (k+1)^2 \qquad \text{Using the definition of sigma notation}$$

$$= \frac{k(k+1)(2k+1)}{6} + (k+1)^2 \qquad \text{By the assumption } S_k$$

$$= (k+1)\left[\frac{k(2k+1)}{6} + (k+1)\right] \qquad \text{Factoring}$$

$$= (k+1)\left[\frac{k(2k+1)}{6} + \frac{6(k+1)}{6}\right]$$

$$= (k+1)\left[\frac{2k^2 + 7k + 6}{6}\right]$$

$$= \frac{(k+1)(k+2)(2k+3)}{6}.$$

Our proof is complete.

## M. Scott Peck and Golf

M. Scott Peck is one of my favorite authors. He is most famous for his three "Road Less Traveled" books, but most recently he wrote a fascinating book entitled *Golf and the Spirit,* in which he draws a parallel of golf to life: "Life is tough! Golf is tough!" Peck has a trigonometry application in the book:

REFERENCES

Peck, M. S., *Golf and the Spirit,* New York: Three Rivers Press, 1999, pp. 54–55.

*Deviate 5° from your aiming point on a 150-yd shot, and your ball will land approximately **20 yd** to the side of where you wanted it to. Do the same on a 300-yd shot, and it will be **40 yd** off target. Twenty yards may well be in the range of safety; 40 yd probably won't. This principle not infrequently allows a mediocre, short-hitting golfer like myself to score better than a long hitter.*

I agree with the philosophical golf assertion in the statement, but Peck's mathematics is bit off. The numbers should be about **13 yd** and **26 yd**, respectively. The math follows.

The solution assumes that when you move 5°, you form a right angle. This means that the deviated drive is a bit longer than the original. Then to compute the deviation *c*, you use some trig as follows:

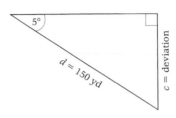

$$\sin 5° = \frac{\text{opposite}}{\text{hypotenuse}} = \frac{c}{150 \text{ yd}}$$

$$\Rightarrow c = 150(\sin 5°) \approx 13.07 \text{ yd.}$$

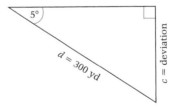

$$\sin 5° = \frac{\text{opposite}}{\text{hypotenuse}} = \frac{c}{300 \text{ yd}}$$

$$\Rightarrow c = 300(\sin 5°) \approx 26.15 \text{ yd.}$$

I wrote Peck a glowing letter about his book and told him about the correction, and he graciously responded. He is a truly fine human being and author, and I urge you to make a point of enjoying his books!

### Pursuing Further

*Truffle Ball Problem.* Suppose the layers in our stacks are triangular, as in this stack of truffle balls. We look at each successive layer from the top down. Look for a pattern.

| 1 ball | 3 balls | 6 balls | 10 balls | 15 balls |
| Top | | | | Bottom |

1. Conjecture a formula for the sum of $n$ of layers of balls in the stack.
2. Prove the following by mathematical induction: For every natural number $n$,

$$1 + 3 + 6 + 10 + \cdots + \frac{n(n+1)}{2} = \frac{n(n+1)(n+2)}{6}.$$

3. Using the results of items (1) and (2) determine a formula, as a polynomial function $T(x)$, for the number of balls in a stack of $x$ triangular balls
4. *Truffle Balls in a Pile.* Use the formula from item (3) to find the number of truffle balls in the stack above.
5. The volume of a sphere of radius $r$ is given by the function

$$V(r) = \frac{4}{3}\pi r^3,$$

where $\pi$ can be approximated as 3.14

Chocolate Heaven has a window display of truffles piled in a triangular pyramid formation 5 layers deep. If the diameter of each truffle is 3 cm, find the volume of chocolate in the display.

## BOWLING AND MATHEMATICS

In contrast to my golf skills, I am a good bowler. Over the past few years my average has ranged from 192 to 206, which is the result of a lot of hard work and practice. I am a bowling fan, keeping up with the pro bowlers on television and attending tournaments. I watched them all—from Johnny King, who bowled with a cigar in his mouth, to Buzz Fazio and Andy Varipapa on black-and-white TV, to Billy Welu, Earl Anthony, and Mike Aulby on the ABC telecasts of the Pro Bowlers Tour. A fan to this day, I enjoy the exploits of Walter Ray Williams, Jr., Norm Duke, Chris Barnes, Parker Bohn III, and big "twisters" like Robert Smith, Bob Learn, Jr., and Rudy (Revs) Kasimakis.

Watching the pros on TV has inspired many people to try to join the Professional Bowlers Association. When I finally completed my education and became established as a math professor, I decided to take the time for lots of bowling instruction and really try to improve. I venture to say that nobody has taken more bowling lessons than I have. In addition to hundreds of lessons here in Indianapolis, I have traveled to Chicago to learn from Tom Kouros, who used to coach the pros on the bowling circuit, to Denver for lessons with Don Johnson, who holds 26 professional bowling titles, and to Las Vegas for lessons with noted instructor Ron DeGroat. Some of it soaks in, and I am eager to learn, but I never get to the level I aspire.

In 1985, at age 44, I earned membership in the Professional Bowlers Association. This was a childhood dream come true, but I was too old and too unathletic to ever consider making bowling my vocation. My dream was just to join and bowl some tournaments but, although I cashed a couple of times, I was never a threat to the stars on the PBA Tour. Once in the PBA National in Toledo I bowled a 268. All-star Gary Dickinson was bowling on the adjacent

REFERENCES
"Bowling and Mathematics," *The COMAP Journal*, Vol. 5, No. 4, 1984, pp. 405–430.
Bittinger, M. L., "A Spare System That Works," *Bowlers Journal*, Vol. 73, No. 1: 122–123, 1986.

lanes and after my 268, he said, "Nice game, Marv!" His words were enough to make this math educator's day!

Some years later in another PBA event, I was bowling with PBA stars John Gant and George Branham. I got strikes on the first 11 of 12 frames. The best score you can get is 300 with strikes in all 12 frames. My knees were shaking on the twelfth shot and I got only 7 pins. After the 297, Branham said to me, "Nice going, I thought you had the 300!" Comments like these from Branham and Dickinson were my PBA "trophies." My lifetime earnings in the PBA are $430.

As with golf and the other applications in this book, it was inevitable that I would find math applications in bowling. I found so many that I eventually wrote a journal article, "Bowling and Mathematics," and a highly streamlined, low-math version of the article published in the Bowler's Journal. I also gave talks on bowling and mathematics at professional meetings. Space and the extensive technical aspects of the game prevent me from reprinting the entire article here. If you would like a reprint write me

c/o Mathematics Editor
Addison-Wesley
75 Arlington St, Suite 300
Boston, MA 02116.

Following is a brief summary of some of my results.

### Standard Deviations

We know that the bowling averages of pro bowlers are usually higher than averages of amateurs. But how do the standard deviations of the pros compare to amateurs? It was my guess that there would be no difference, but I was wrong.

Most people have no concept of standard deviation. I like to use the notion of a "spread" number to describe the concept. Standard deviations measure the "spread" of data about the mean in a distribution, as seen in Figure 5.

FIGURE 5

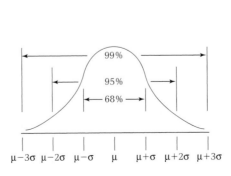

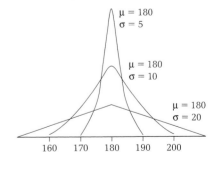

*The standard deviation $\sigma$ indicates the "spread" about the mean, $\mu$.*

To test my hypothesis that pros and amateurs had about the same standard deviations in their bowling, I gathered data from PBA tournaments and that of some of my amateur friends in leagues. It is displayed in the following table.

| Bowlers | $\mu$ | $\sigma$ | $1/n$ represents 1 chance in $n$ of scoring 290 or higher |
|---|---|---|---|
| **Pros** | | | |
| Earl Anthony | 229.8 | 29.75 | 1/46 |
| Mike Aulby | 203.6 | 28.05 | 1/719 |
| Dave Husted | 210.3 | 27.49 | 1/518 |
| Randy Stoughton | 207.7 | 28.59 | 1/503 |
| Gary Dickinson | 202.3 | 34.30 | 1/191 |
| Steve Cook | 213.6 | 29.61 | 1/202 |
| Bob Handley | 207.8 | 29.27 | 1/403 |
| Wayne Webb | 205.8 | 28.99 | 1/535 |
| Tom Milton | 209.1 | 32.84 | 1/144 |
| Marshall Holman | 207.9 | 24.15 | 1/571 |
| Mark Roth | 213.5 | 25.32 | 1/901 |
| **Amateurs** | | | |
| MB | 193.8 | 26.2 | 1/8000 |
| DW | 204.5 | 29.4 | 1/535 |
| JB | 176.2 | 16.9 | 1/100,000 |
| RG | 196.5 | 33.5 | 1/376 |
| BC | 186.9 | 27.6 | 1/10,695 |
| GK | 201.8 | 29.5 | 1/784 |
| CA | 178.9 | 21.5 | 1/100,000 |
| OF | 171.6 | 20.3 | 1/100,000 |
| SF | 211.4 | 28.2 | 1/382 |
| FR | 165.8 | 23.9 | 1/16,667 |
| DH | 156.7 | 21.6 | 1/100,000 |

*Mike Aulby, my favorite bowler, and one of the finest persons I know.*

As the table suggests, I was indeed wrong about my hypothesis. I found that the standard deviations of pros are much higher than that of amateurs. This conclusion was carefully supported statistically with the aid of Professor Robert Kleyle in the Indiana University-Purdue University at Indianapolis math department. Further details can be found in the COMAP article cited earlier.

## Spare Conversion

There are 10 pins to knock down in bowling. You score a "strike" when you knock them all down. When some pins are left and you knock those all down on your second ball, you score a "spare." In Figure 6, we see some aspects of a bowling lane (not drawn to scale).

FIGURE 6

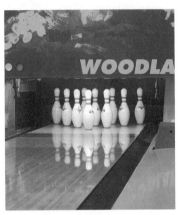

*Each similar set of three pins is an equilateral triangle 12 in., or 11.1306 bds on a side.*

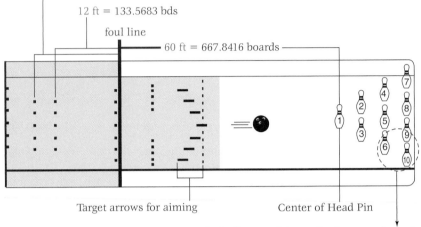

15 ft = 166.9604 bds

Marks on the approach for feet location

12 ft = 133.5683 bds

foul line

60 ft = 667.8416 boards

Target arrows for aiming          Center of Head Pin

Each similar set of three pins is an equilateral triangle 12 in., or 11.1306 bds on a side.

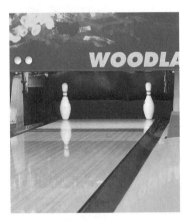

*The 7–10 split.*

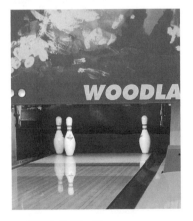

*The 4–7–10 split.*

A bowling lane is about 41.5" wide, ±0.5", and is constructed of long strips of laminated wood. Each strip is called a "board," the width of each board being about 1.0781". From the foul line to the center of the headpin, or 1-pin, is about 60 ft. It is common for a bowler to stand a certain distance from the foul line and aim at a target for his first ball. If he does not get a strike, it is typical to move a certain number of boards and aim at the same target to convert the spare.

In one of my first bowling lessons I was taught to move my feet three boards to the right when converting the 2-pin, six boards for the 4-pin, and nine boards for the 7-pin. These moves never seemed to be enough for me. Four boards worked better for the 2-pin, ten boards for the 4-pin, and eleven boards for the 7-pin.

As I gained experience and watched other bowlers, the mathematician in me led to the discovery of at least three variables that accounted for the discrepancy between the 3–6–9 system I was taught and what I learned from trial and error.

First, the arrows are not the same distance from the foul line. Second, the pins are not the same distance from the foul line. Third, and foremost, bowlers stand on the approach at different distances from the foul line.

Using trigonometry and a lengthy computer program, I set out to revise the thinking of simply moving a fixed number of boards. Since it would be too lengthy to include the development here, I have abbreviated the results in the following table.

If you have a TI-83 or TI-83+, you can program your calculator to do the conversion for you. The following program was created by my dear friend, fellow math professor, and PBA bowler, Dr. Edward A. Zeidman. Program your calculator with this and take it to the lanes to wow your friends.

Move Your Feet Spare Systems

| Strike Target | Approach Location (ft) | Feet Should Move to Convert (bds) | | | $d$ (board) |
|---|---|---|---|---|---|
| | | 2-Pin | 4-Pin | 7-Pin | |
| First Arrow | 10 | 4 | 7 | 9 | 2.0 |
| | 12 | $4\frac{1}{2}$ | $7\frac{1}{2}$ | 10 | 1.8 |
| | 15 | 5 | 8 | 11 | 1.6 |
| Second Arrow | 10 | $4\frac{1}{2}$ | 7 | $9\frac{1}{2}$ | 1.9 |
| | 12 | $4\frac{1}{2}$ | $7\frac{1}{2}$ | $10\frac{1}{2}$ | 1.8 |
| | 15 | 5 | 8 | $11\frac{1}{2}$ | 1.6 |
| Third Arrow | 10 | $4\frac{1}{2}$ | $7\frac{1}{2}$ | 10 | 1.8 |
| | 12 | 5 | 8 | 11 | 1.7 |
| | 15 | $5\frac{1}{2}$ | 9 | 12 | 1.5 |
| Fourth Arrow | 10 | $4\frac{1}{2}$ | $7\frac{1}{2}$ | $10\frac{1}{2}$ | 1.8 |
| | 12 | 5 | 8 | 11 | 1.7 |
| | 15 | $5\frac{1}{2}$ | 9 | $12\frac{1}{2}$ | 1.5 |

## START OF PROGRAM

```
ClrHome:Radian
Disp "START POSITION"
Input "IN FEET=",S:S*12/1.0781→S
Input "TARGET BOARD AT ARROW",Y

If Y ≤ 20:(16.7/15)(Y-20)+170.7→X
If Y ≥ 20:"(16.7/15)(Y-20)+170.7→X

√((X-672.6613)²-(Y-17.2173)²→E
S+X→F
Disp "X1=",X
Disp "Y1=",Y

Disp "E=",E
Disp "F=",F

For(A,1,12)

A→L₁(A)
tan⁻¹(A/F)→D
D*180/π→L₃(A)
E*sin(D)/sin((π-D)/2)→C
C→L₂(A)
```

```
tan⁻¹((Y-17.2173)/(X-672.6613)))→H
E*cos(D+H)+X→T
E*sin(D+H)+Y→R
(R-Y)/(T-X)*(677.48-X)+Y→L₄(A)
(R-Y)/(T-X)*(687.12-X)+Y→L₅(A)
(R-Y)/(T-X)*(696.76-X)+Y→L₆(A)


End


Disp "PRESS [STAT] 1"
                    END of Program
```

Doing the research on the standard deviations of professional bowlers and developing this spare conversion program was an **MPX** that occurred while I was enjoying the application of mathematics to one of my favorite games.

**MPX** MUSIC: "Axel F" by Walter Faltermeyer; on the sound track to the movie *Beverly Hills Cop*, © 1984 MCA Records. This used to be the theme music of the PBA at its tournaments.

Bowling suffers from a poor image. People think of it as a recreation rather than a sport. Believe me, lots of knowledge of steady-state physics is involved in the art of drilling the finger holes in bowling balls. Balls can weigh more on the top than on the bottom and more on one side than the other and remain within legal guidelines established by the American Bowling Congress. A knowledgeable bowler can adjust these weights to make his ball hook earlier or later. Thinking this out can be a daunting task.

The following reprint of an article about my career and hobbies appeared in the January 1997 issue of the *Bowler's Journal*. The editor of the *Bowler's Journal*, Jim Dressel, commissioned my good friend and radio newsman for local station WIBC, Dennis Bergendorf, to write this article. I was thrilled that Dennis seemed to capture my soul in the article.

## THE ALGORITHMS OF SPORT

### Leave it to a mathematics professor to appreciate the innate appeal of bowling and baseball.

by Dennis W. Bergendorf

Take a quick glance of Marv Bittinger, and it's easy to miss the passion that simmers beneath the surface. Unassuming, soft-spoken, and with skin a shade or two short of a golden glow, he projects a countenance that is decidedly professorial.

He should. He holds a full doctorate in mathematics education, teaches in a giant state university system, and is one of the world's most prolific authors of math textbooks.

*You might call Marv Bittinger a math professor whose passions for bowling and baseball just don't figure. But the good doctor loves both as much as an algorithm.*

He also happens to be a professional bowler. Or, to be a bit more accurate, he holds a PBA Resident II card and has, on a few occasions, crossed with some of the top names in the sport.

That he hasn't exactly whipped those top names hardly makes him a candidate for emotional counseling. Marv Bittinger is living a dream: using his love of teaching (and writing) to finance a passion for bowling . . . and baseball. You see, in addition to hitting a few regionals and a national stop or two, twice a year he tries to hit 85 m.p.h. fastballs thrown by major league pitchers in Los Angeles Dodger fantasy camps.

Growing up in Akron, not much more than a long fly ball from Cleveland, he was a natural Indians fan. In those post-war summers, when "a kid would rather skip a meal than miss an inning of sandlot ball," the diamond ruled. He gave his heart to the game, but alas, the game did not reciprocate. At the tender age of 12, Marv Bittinger was cut. From Little League.

He remained a fan, but with a 60-by-60 foot diamond-shaped hole torn from his soul, it wasn't quite the same. He turned his attention to other sports, finally giving a shot to bowling. Soon, he was hooked on that sport and would spend weekdays dreaming of the next Saturday morning junior league session, despite the three-mile walk to the bowling center.

### Hello, Priorities; Goodbye, Bowling Dreams

Had it been a few years earlier, he might have become a devotee of the professional game; but the Tour was in its infancy, and the Firestone had yet to hold its inaugural Dinner of Champions. Still, he was aware of and impressed with the greats who punctuated the Akron scene (in fact, Hall of Famer Dick Hoover drilled Marv's first ball).

His average grew with his physical stature, until he was rolling at a 160 clip. But following high school, he realized education was his future, not sports. The result: "I let bowling slide."

Moving to Indiana in 1968 to complete his doctorate at Purdue University, Bittinger saw his future take three significant turns. Bittinger met Dr. Mervin Keedy, who asked his help in completing a mathematics text book. He took up adult softball. And he dusted off the old bowling shoes, joining a winter league.

It's safe to say that, for the most part, he picked up where he left off, which means he didn't exactly terrorize Mike Aulby and Company. Languishing in the 160-range, Bittinger began seeing the game as a matter of logic. He felt he could get within shouting distance of those guys on the Saturday telecasts if he got some instruction.

Soon he was listening to Indianapolis professionals John Combs and Randy Stoughton, and his game started a slow climb to respectability. From 165 in 1981, through the 180s, until, in 1985, he cracked the 190 barrier and, in celebration, treated himself to a PBA card.

He has vivid memories of his first pro tournament, the True Value Open at nearby Woodland Bowl. Green as a pool table, he felt the sheer intimidation of sharing a paddock with bowling's superstars. (He crossed with Jeff "The White Whale" Mattingly, one of the first monster crankers. Also on his pair was Bob Benoit, but "That was before his 300 game, and he wasn't much of a name.")

The other impression was the toughness of the condition and the way the stars ate it up. Darryl Bower won, averaging close to 230. Bittinger finished his 18 games more than 230 pins under.

But during this period, his love of baseball was rekindled, in the form of softball, when he put together a team called "The Exponents." He was soon rounding the bases about as well as most 40-somethings, and a business associate was bugging him about attending one of the Major League Baseball fantasy camps that were springing up in the Sunbelt. So in 1989, with the winds of early winter howling through the Indiana corn stubble, the two headed off to Vero Beach, Fla., to rub elbows and hit fungoes with members of the Dodgers.

## Dr. Baseball, I Presume

That first session was every bit as intimidating as the True Value Open. Between the lines were Tommy LaSorda and all-stars Carl Erskine, Reggie Smith and Dusty Baker (acknowledged as one of the great hitting instructors of the day).

About the same time, with his own textbook-writing career now well established, he ran into Jeff Mercer, the owner of a successful batting cage facility. Mercer had some ideas about the science of hitting a baseball and had himself written a manuscript. On the next trip south, Bittinger showed the book to Baker. Baker liked it, added some advanced material, and "You Can Teach Hitting" was born.

Fortune again smiled on a Bittinger enterprise, as the book was published in late 1993, and Baker was named manager of the San Francisco Giants two months later. So far, the book has sold more than 22,000 copies, considered a decent run for a specialty book of its type.

In terms of numbers, though, Bittinger's primary vocation is considerably more impressive. His 130 mathematical titles qualifies him for a McDonald's-like proclamation of success: over seven million sold.

Unlike hit-or-miss popular markets, textbooks sell to a rather captive audience. Once a book is adopted by a college professor, it's a mandatory purchase for students (at larger schools, hundreds pony up as much as $50 to buy a new copy). And while the royalty scale varies (Bittinger never works on monetary advances), his 50,000 or so sales a year translates into an income that makes the payments on a home in the suburb of Carmel and readily funds two fantasy camps and a couple of regionals a year, even if he rarely goes to the pay window (in 17 events, he's cashed twice, for a total of $430).

## Making the (Math) Road Less Bumpy

Bittinger might be a shrewd businessman, but the money's not really the issue (in fact, wife Elaine balances the checkbook). But ask him about teaching mathematics and watch his face light up. "My love is to convey information to people who don't appreciate it initially," he says.

"In math, students have to get from point A to point B. If I can make the road easier, I've succeeded. I can't make the math easier, but I can make the road less bumpy."

To turn Bittinger's smile upside down, mention the state of America's math knowledge and understanding. When he started, it was accepted as a discipline of life, a skill to be mastered. These days, new acquaintances are apt to mention that they are terrible at math and don't care much for it. "People see math teachers as something akin to dentists—someone who brings pain."

He cites ex-Washington Redskins star Dexter Manley, recently released from prison, who burst into tears when confessing on national television that he couldn't read. Bittinger laments that "nobody cries when they admit they can't do math."

That's one reason why Bittinger concentrates on remedial text books. His colorful seventh edition of "Basic Mathematics" is aimed at a generation raised on MTV and Saturday morning cartoons. It's filled with pictures, charts and exercises to keep the imagination, as well as the intellect, occupied.

As for channeling to bowling his wealth of mathematics knowledge, Bittinger is a bit hesitant. "It can improve spare-making, but striking has become more a function of physics," a discipline he concedes to other experts.

He has written a guide to spares, published in *Bowlers Journal* in 1986. Earlier, a professional journal accepted a remarkably deep treatise on probabilities and formulas for spare-shooting.

But for 1997, the writing will focus on remedial math and another edition in his popular "Math Max" series.

The passion will be directed toward Dodger fantasy camps and crossing lanes with the greatest bowlers in the world.

### *Pursuing Further*

The following math problem appeared in the December 1982 edition of the *Bowler's Journal.* It had been in a recent *Games Magazine* as an article entitled "Alley 1." This puzzle generated a great deal of interest. Free subscriptions were given away for the first five correct solutions. The fact that the magazine received more than 400 correct answers was evidence to me that bowlers are indeed quite smart. Try the problem by yourself or in the classroom. It is rich in the use of many kinds of mathematics.

Art, Bob, Carl, Denny, and Fred are on the same bowling team. They are all being factual in the following statements regarding the last game they bowled that night.

| | |
|---|---|
| Art: | My score was a prime number. Fred finished third. |
| Bob: | None of us bowled a score over 200. |
| Carl: | Art beat me by exactly 23 pins. Denny's score was divisible by 10. |
| Denny: | The sum of our five scores was exactly 885 pins. Bob's score was divisible by 8. |
| Fred: | Art beat Bob by less than 10 pins. Denny beat Bob by exactly 14 pins. |

Determine each bowler's score in the game.

# Baseball and Mathematics

# 3

**INTRODUCTION**

*It [baseball] breaks your heart. It is designed to break your heart. The game begins in the spring when everything else begins again, and it blossoms in the summer, filling the afternoons and evenings, and then as soon as the chill rains come, it stops and leaves you to face the fall alone.*

—A. BARTLETT GIAMATTI,
former commissioner of
Major League Baseball

*Dusty Baker in Cubs uniform.*

There are a number of things in life that I love dearly, but I'd say the top five are:

1. My wife and family
2. Mathematics
3. Baseball
4. Hiking in Moab, Utah
5. Bowling

My wife agrees with the list, but not necessarily the order, wondering if baseball is really number 1. In truth, I can walk out onto a baseball field amid the red dirt, the white chalklines, and the freshly mown grass, and just feel the goodness of life. The thought quickly comes to mind, "Where are the guys? Let's have a game!" I am not a great, not even a good, player but my passion for the game is great!

Upon seeing his picture, you are probably asking what is this textbook author doing with the former manager of the San Francisco Giants and present manager of the Chicago Cubs, Dusty Baker? And what does his picture have to do with mathematics? The answer to these questions is a fantasy come true. It is a bizarre story that springs from my love of baseball and mathematics.

I grew up in a neighborhood with numerous empty lots. We spent most of our summer days on those lots playing ball from morning to night. We used any kind of ball we could find—new, taped, hardball, or softball; and any kind of bat, mostly filled with glue and nails and covered with tape. It saddens me now as I drive by a school yard near my house and see two beautiful ballfields and backstops that are rarely used. How my baseball buddies and I would have loved to play on those fields! As most of my friends and family will tell you, I love to eat. But, when my grandmother, a tremendous cook, would call me to come home for supper and we were playing baseball, I never wanted to go.

*Dusty and Marv at Reds–Cubs game in 2003.*

I love to ask the following question, "Can you name two people ever cut from a Little League baseball team?" Ironically, I know two such guys. One was my dear friend and present manager of the Chicago Cubs, Dusty Baker, cut by his father because he had a bad attitude. The other was yours truly at the age of 12 for lack of playing ability. Little League was organized late in Akron compared to other cities, so there were only a few teams with too many kids to fill them. So, they held tryouts. I went to a tryout with great expectations, only to be cut—a devastating experience to a 12-year-old boy who just loved to play ball. In my mind, my baseball days were over. For the next 20 years, my baseball involvement was limited to attending Cleveland Indians games and listening to Jimmy Dudley broadcast the games on WAKR.

My early memories of TV were rooting for the Brooklyn Dodgers to beat the hated New York Yankees in the World Series. I detested the Yankees because, other than in 1954, my Indians placed second to them every year. I remember all too well watching the Indians play the Giants in the 1954 World Series. I know right where I was sitting when my heart was torn out as Willie Mays made that incredible catch in centerfield off the bat of Vic Wertz. What 13-year-old kid sitting in front of the TV set would have dreamed that someday he would write a book with the manager of those same Giants, and actually shake the hand of Willie Mays? Not me, not me, NOT ME!!

Each of us goes through the process of discerning our strengths and weaknesses. I realized very early in life that I wasn't going to be a professional baseball player. I moved on to academic pursuits, where I seemed to have more God-given talents. But at the age of 32, when I had my education

*Willie Mays making that catch in the 1954 World Series.*

completed and my profession as a math educator well under way, I started playing and coaching some slow-pitch softball teams.

Some years later I came to know Barton (Bart) L. Kaufman. Bart was an outstanding athlete, playing outfield for Indiana University. In 1961 he finished second in the Big Ten batting title race, first place going to Bill Freehan, who later caught for the Detroit Tigers. Bart could hit, and he still can. He calls me "Right-hander" from our days playing softball together.

Bart was recommended to me as a financial advisor around 1975. I was drawn to him because of his tremendous intelligence, outstanding character, and knowledge of life insurance and estate planning. But I must admit the crowning part of my decision to work with him was his passion for sports and especially the Cincinnati Reds. Our relationship through the years has evolved from business acquaintances to very close friends. It would take another chapter in this book to relate all the contributions Bart has made to my life; suffice it to say they approach infinity as a limit.

Bart's father, wanting to give him every advantage, sent his teenage son to summer instructional camp at the Dodgers spring training complex in Vero Beach, Florida. At that same complex, in the mid-eighties, the Dodgers started their Adult Baseball Fantasy Camps, offered to anyone at least 30 years old who is willing to pay the admission. Campers go to the Vero Beach spring training site, Dodgertown, take instruction and play a 9-inning baseball game every day for a week. They sleep in the same beds as the Dodger players and eat the same food. The camps are coached by former Dodger stars such as Carl Erskine, Reggie Smith, Steve Garvey, Jerry Ruess, Jeff Torborg, and Tommy Davis.

Bart attended these camps and, knowing my love of the game, urged, if not begged, me to attend with him. But I continually refused, explaining that I had only one good eye (the other is what used to be called a "lazy eye" so I have no depth perception) and that I was so lacking in talent that I had even gotten cut in Little League. However, after he told me Jane Fonda attended one year and that he needed a roommate, I finally went in 1989. I loved it so much that I have gone every year since, and sometimes twice per year.

Dodger Camp is truly baseball heaven. We usually have instruction in the mornings and play a nine-inning baseball game in the afternoons. In the evenings we meet together for supper and awards, and are regaled with war stories from the former players. At the end of the week, the campers play a game against the former players—they slaughter us, but that makes for more good stories. I will always remember getting base hits off Jerry Reuss and Ron Perranoski. I also faced the great Hall-of-Famer, Bob Gibson, who struck me out. With each pitch I was scared to death that I might get hit and die! He actually just lobbed up the pitches, but that did not relieve my fears.

It was my first time at camp in the fall of 1989 when I met Dusty Baker, a former Dodgers star, who was then batting coach for the San Francisco Giants. Dusty gave me two 45-minute hitting lessons. What a thrill for a half-blind guy who got cut from Little League!

*Bart and Marv in Dodger uniforms.*

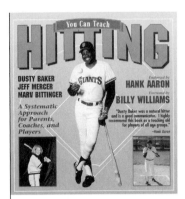

*You Can Teach Hitting.*

*Dusty Baker book and videos. There is also a CD-ROM.* For more information on the Dusty Baker Hitting Products visit www.dustybaker.com.

I had a great time and became obsessed with these camps, looking forward to them each year with a goal of improving my hitting skills. When I was looking for batting lessons one year I met Jeff Mercer, former assistant coach at Indiana University, who was running an indoor baseball facility here in Indianapolis. Jeff was an exceptional instructor and was willing to help an old guy with one eye and poor athletic ability. When he learned that I was an author, Jeff told me he had always wanted to write a book on hitting and asked how he might go about it. Believe me, I have had many discussions with people who want to write; I usually suggest they write an outline, a proposal, and maybe a sample chapter. Most of the time, the idea blows into the wind. But that wasn't true in Jeff's case; he came back to me two weeks later with an outline and a sample chapter—and it was good!

I was determined to get this book published but felt we needed someone with a national baseball reputation as a co-author. So I contacted Dusty Baker and met him for lunch in Cincinnati when the Giants were in town playing the Reds. He liked the idea of writing a book and producing a set of instructional videos, and later a CD-ROM.

*A college baseball coach, Jeff Mercer, a professional baseball manager, Dusty Baker, and a professional mathematician, Marv Bittinger ready to play ball! No matter the job, they possess the same love of the game.*

By now you are probably wondering how all the preceding leads to math and baseball. At the end of our lunch, when it was time for Dusty to go to the ballpark, he mentioned that since we were going to write a book together, I could go to the park with him and go into the locker room. Then my adrenaline started flowing like never in my life—I was going to the San Francisco Giants locker room!

We took a cab to the stadium, drove down underneath to the visitors entrance, and went into the locker room with Will Clark, Matt Williams, Kevin Mitchell, Robby Thompson, and all the other players. Wow! Dusty introduced me to some people at a table nearby and I sat down with Wendell Kim, third base coach, Bob Lillis, bench coach for then-manager Roger Craig, and Syd Thrift, who was employed at that time as a consultant to the Giants.

Syd Thrift may not be a familiar name but with Barry Shapiro he wrote the book *The Game According to Syd*. When he was general Manager of the Pittsburgh Pirates, Syd studied scientific ways his teams could get an edge on the competition.

Dr. John Nash Ott was a team consultant to the Pirates. Ott was an expert on environmental health and light research. Ott found that light influences the pituitary gland, which controls the endocrine system, which in turn governs the production and release of body hormones, which control body chemistry. Ott also discovered that a player's performance will improve if the underside of the bill of the baseball cap is gray instead of green. Today all Major League Baseball teams use gray under the bills of their caps.

Another edge was Syd's knowledge that a four-seam fastball is 4 mph faster than a two-seam fastball. This doesn't mean that the hide of a baseball is sewn differently. Rather, it refers to the placement of the pitcher's fingers on the ball in relation to the ball's stitches.

Not being sure how to start a conversation with these men, I said to Syd, "I read in your book that the four-seam fastball is 4 mph faster than the two-seam fastball." Syd's eyes lit up and he asked me, distance-wise, how much faster did one ball get to the plate over the other? Well, I was so excited to be in that locker room that (for once) I did *not* want to do math. I tried to escape the issue by asserting that I needed a calculator. But right away Bob Lillis handed me the calculator-organizer in which he kept his scouting data. I finally got the answer: The two-seam fastball comes in 30 inches behind the four-seam fastball. (We work out the problem on page 65.)

This was the first of my baseball and mathematics applications. Through all the subsequent time I spent with Dusty and his players, many other applications cropped

REFERENCE
Thrift, S., and Shapiro, B. *The Game According to Sid: The Theories and Teachings of Baseball's Leading Innovator,* New York, Simon & Schuster, 1990.

*Two of Dusty Baker's real baseball cards:* (left) *Dusty as a player with the Dodgers in 1983 and* (right) *Dusty when he played with the Giants in 1985.*

up. Believe me, players do math in the dugout! These ideas formed the basis of a math talk that I give at many mathematics education conventions.

### The Strike Zone

Another very elementary application concerns the strike zone.

*The strike zone is an area of about 17 in. by 40 in., depending on the height of the hitter. What is the effect on its area of adding a 2-inch border to the zone?*

The inside, or actual, area is

$$(17 \text{ in.})(40 \text{ in.}) = 680 \text{ in.}^2.$$

When the 2 in. is added, we get a new outside zone,

$$21 \text{ in. by } 44 \text{ in.}$$

The new area is

$$(21 \text{ in.})(44 \text{ in.}) = 924 \text{ in.}^2.$$

The new area has been increased by

$$924 \text{ in.}^2 - 680 \text{ in.}^2, \text{ or } 244 \text{ in.}^2$$

Thus, there is an increase of

$$\frac{244}{680} \approx 0.36 = 36\%.$$

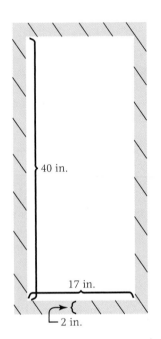

40 in.

17 in.

2 in.

How is this relevant to a major league hitter? I once heard Ted Williams being interviewed by Bob Costas on a radio program. Williams mentioned a theory that he had developed on his own but checked with other baseball players and several umpires: "When a major league hitter swings and misses on strike three, 70% to 80% of the time the pitch was out of the strike zone." I checked with Dusty and he agreed with Williams. In connection with the 2 inches added to the strike zone, the hitter may be allowing the pitcher 36% more area to pitch to. A pitcher and the hitter should be aware of this change. It supports a theory: "Give a pitcher 36% and he will strike you out!"

#### *Pursuing Further*

*Baseball's Strike Zones.* Major league rules define the true strike zone to be the rectangle *ABCD* shown in Figure 1—knees to name between the sides of the plate. Somehow, prior to 2001, the strike zone had evolved to a diminished region, *AQRST*. In 2001 the major leagues mandated that the umpires return to enforcing the true strike zone. Figure 1 represents the zones for a normal-sized player. The height changes with the height of the player.

1. Find the area of the true strike zone, *ABCD*.
2. Find the area of the altered strike zone, *AQRST*.

3. How much larger is the true strike zone than the altered strike zone?

4. By what percentage had the area of the strike zone *ABCD* been diminished when it evolved to *AQRST*?

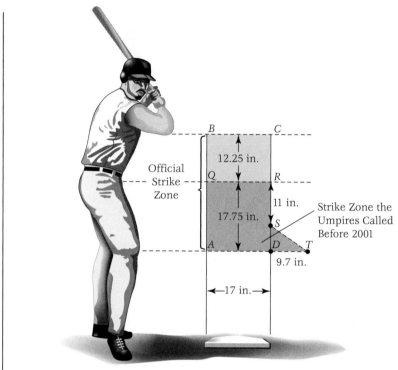

FIGURE 1
*The strike zone according to major league rules.*

REFERENCES
Thrift, S., and Shapiro, B. *The Game According to Sid*. New York, Simon & Schuster, 1990, p. 114.
Baker, D., Mercer, J., and Bittinger, M. *You Can Teach Hitting*, Indianapolis, Bittinger Books, 1993, p. 197. www.dustybaker.com.
Bittinger, M. L., *Calculus and Its Applications*, 8th ed. Boston, Addison-Wesley Longman, 2004, pp. 181–183.

## FOUR SEAMS VS. TWO SEAMS

Here we solve the four seams vs. two seams problem as posed to me in the Giants locker room. According to the results of Thrift's studies, the four-seam fastball travels 4 mph faster than the two-seam fastball. But how much faster, distance-wise, does a four-seam fastball get to the plate than a two-seam fastball? We start by trying to understand the difference between the two kinds of pitches. The answers lie with the seams and the grips.

### Four-Seam Fastball

In Figure 2A, we see how the forefinger and middle finger are placed on the baseball in order to throw a four-seam fastball. The ball is thrown toward the batter as if the hand is pointing toward the hitter. The pitch is released with backspin. With this grip imagine two horizontal seams on this side of the ball and two on the side we can't see. When the ball rotates with its backspin, each rotation creates "four seams." See Figure 2B. Thus, we call this a four-seam fastball. To see this better, hold a baseball as shown in Figure 2A, and imagine rolling it off the tips of your fingers with backspin toward the hitter.

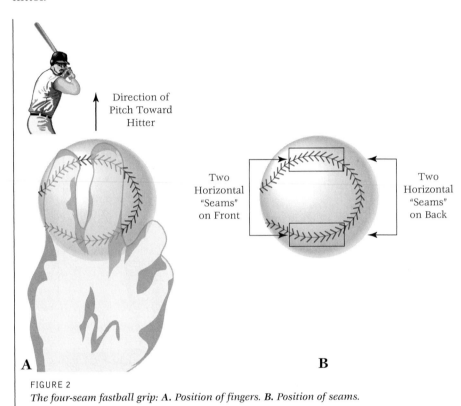

FIGURE 2
*The four-seam fastball grip:* **A.** *Position of fingers.* **B.** *Position of seams.*

### Two-Seam Fastball

In Figure 3A we see how a two-seam fastball is gripped. But, now only two horizontal seams go against the wind with the backspin; hence, the term for such a pitch is "two-seam" fastball. See Figure 3B. The four-seam grip is compared to the two-seam grip in Figure 4.

Why does one ball travel faster than the other? The simplest answer comes by proving it through repetitive testing with throwing machines. But let us try to examine the situation using some physics. Look at the drawings in Figure 5. When the balls are thrown, a turbulent wake is created behind each ball. The wake from a four-seam pitch is greater than the wake from a two-seam pitch. One conclusion we can draw is that the wake tends to create a molecular "push," causing the ball with the most wake, or push, to go faster. I think of the seams creating a vacuum around the ball. There is more vacuum with the four-seam fastball than with the two-seam fastball. Thus, the four-seam fastball goes faster.

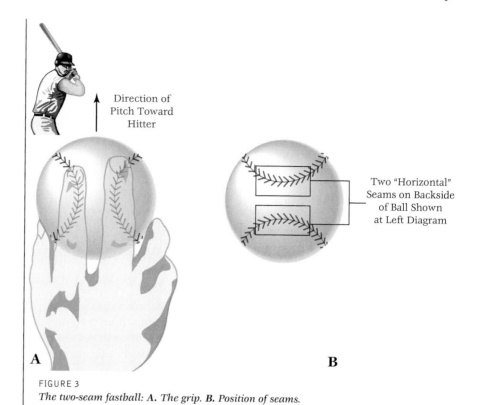

FIGURE 3
*The two-seam fastball:* **A.** *The grip.* **B.** *Position of seams.*

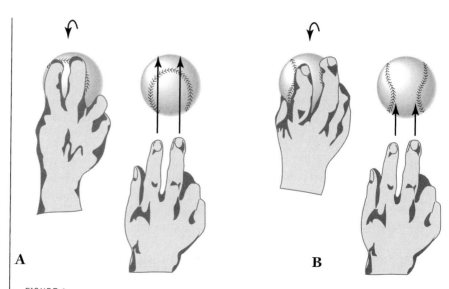

FIGURE 4
*Comparision of fastball grips:* **A.** *The four-seam fastball.* **B.** *The two-seam fastball.*

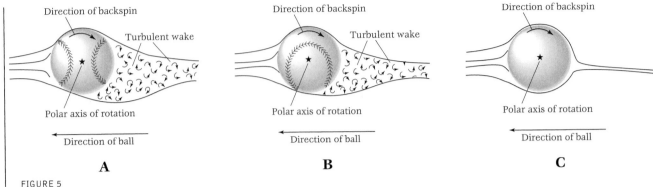

FIGURE 5
*Airflow over a fastball:* **A.** *Four-seam fastball.* **B.** *Two-seam fastball.* **C.** *No seams.*

REFERENCE
Adair, R. K., *The Physics of Baseball*. New York, Harper & Row, 1990. I am indebted to Prof. Adair for the information in his book, in letters, and in several telephone conversations. Adair is a physics professor at Yale University and was appointed Physicist of the National League from 1987 to 1989, by his dear friend Bart Giamatti, who at that time was Commissioner of Major League Baseball.

If we had no seams on the ball, as in Figure 5C, the air would be right up against the ball, causing more friction and making it travel slower. This idea is corroborated by the Professional Golfer's Association: A golf ball that goes 260 yd with dimples, which act like seams causing turbulence, would travel only 150 yd if it had no dimples. According to Adair, if you had a hundred highly qualified physicists in one room, you could not get them to agree on the theoretical reason why one ball travels faster. Thus, the "proof is in the pudding," so to speak, by doing the repetitive testing with a throwing machine.

Now let's look at the mathematics and solve the problem. Major league fastballs travel anywhere from 85 mph to 100 mph. Suppose a ball thrown with a two-seam grip travels at a speed of 86 mph. If the same ball is thrown with a four-seam grip, it will travel 90 mph:

Two-seam: 86 mph

Four-seam: 90 mph

The pitcher's rubber is 60 ft, 6 in. from home plate. When a pitcher delivers the ball, he normally leans toward the plate, shortening the distance the ball travels to about 56 ft because it does not actually leave the pitcher's hand until he drives his body off the mound toward the hitter. At 90 mph, how long does it take, in seconds, for the pitch to reach the plate? We use the distance formula, substituting 56 ft for distance, $d$, and 90 mph for the speed, or rate, $r$:

$$d = t$$

$$56 \text{ ft} = 90 \, \frac{\text{mi}}{\text{hr}} \cdot t \quad \text{Substituting}$$

Next we solve for $t$, in seconds, by multiplying on both sides by $\dfrac{1 \text{ hr}}{90 \text{ mi}}$, and convert the result to seconds by making the appropriate unit conversions:

$$56 \text{ ft} = 90 \frac{\text{mi}}{\text{hr}} \cdot t$$

$$56 \text{ ft} \cdot \frac{1 \text{ hr}}{90 \text{ mi}} \cdot \frac{1 \text{ mi}}{5280 \text{ ft}} \cdot \frac{60 \text{ min}}{1 \text{ hr}} \cdot \frac{60 \text{ sec}}{1 \text{ min}} = t.$$

Simplifying, we get

$$t = 0.424242\ldots \approx 0.42 \text{ sec.}$$

Thus, the four-seam fastball travels 56 ft, or 672 in., in 0.42 seconds. The two-seam travels how far in 0.42 sec? To find out, we use the distance formula and substitute $86 \dfrac{\text{mi}}{\text{hr}}$ for $r$, and 0.4242 for $t$. Then we make unit changes to get the distance in inches.

$$d = t$$

$$d = 86 \frac{\text{mi}}{\text{hr}} \cdot (0.4242 \text{ sec})$$

$$d = 86 \frac{\text{mi}}{\text{hr}} \cdot \frac{5280 \text{ ft}}{1 \text{ mi}} \cdot \frac{12 \text{ in.}}{1 \text{ ft}} \cdot \frac{1 \text{ hr}}{60 \text{ min}} \cdot \frac{1 \text{ min}}{60 \text{ sec}} \cdot (0.4242 \text{ sec})$$

$$d = 642.07 \text{ in.}$$

Subtracting the two-seam distance, 642.07 in., from the four-seam distance, 672 in., we see that the two-seam fastball is about 30 in. behind the four-seam fastball. These computations confirm the result I determined in the Giants locker room in Cincinnati.

Why were these coaches so interested in the answer to the question? Getting "jammed," meaning not getting the bat off your shoulder or barely starting a swing, is the ultimate embarrassment. Of course, a pitcher loves to "jam" a batter in this way. The 30 inches can mean all the difference in the world to a hitter and a pitcher.

There are two sidelights to this rather simple application of mathematics. One occurred when I saw Bob Lillis, then Dusty's bench coach, at a Giants game a year or two after that initial locker-room challenge. He again asked and I confirmed the 30-inch difference. Sometime later I also shared this information with Tom Beyers, then minor batting coach for the Los Angeles Dodgers, and Dave Wallace, who was the pitching coach for the major league Dodgers and is now the pitching coach for the Boston Red Sox. Both were fascinated with the results.

This connection between mathematics and the speeds of pitches was a genuine ▮MPX▮. The enthusiasm of relating the results to professional hitting and pitching coaches was in some small way a contribution to baseball from a guy who got cut from Little League and grew up to be a mathematician.

**MPX** MUSIC: "Centerfield," (John Fogerty, arr. Steven Reineke) Wanaha Music Company (ASCAP). Tom Wopat, vocals, Timothy Berens and Jerry Kimbaugh, guitars. Track 1 on the CD *Play Ball!* by the Cincinnati Pops, directed by Erich Kunzel.
MOVIE: *Bull Durham* (1988) starring Kevin Costner and Susan Sarandon; my all-time favorite baseball movie. When Annie Savoy (Susan Sarandon) said that on Sunday morning she went to church at the cathedral of baseball, I was a goner.

### Pursuing Further

1. Consider: Four-seam: 86 mph; two-seam: 82 mph.
   How much faster, distance-wise, does the four-seam fastball get to home plate than the two-seam fastball?
2. Consider: Four-seam: 94 mph; two-seam: 90 mph.
   How much faster, distance-wise, does the four-seam fastball get to home plate than the two-seam fastball?
3. Compare your results from item 2 with what we found in this section. Does it matter what speed we start with on the four-seam fastball?
4. This alternative problem occurred to me in June of 2003 when I went to a Reds vs Cubs game with Dusty in Cincinnati and was able to be on the field for batting practice. The coaches typically throw batting practice to the players, but at slower speeds than most active major league pitchers. This is because the coaches are older and many were not pitchers in their playing days. To compensate for their slowness, teams move the pitching rubber 10 ft closer to the plate to a distance of about 50 ft, 6 in., the goal being to emulate the faster pitches they see in real games. Wendell Kim, one of Dusty's coaches for the Cubs, told me he estimated that he threw 55-mph pitches to the players in batting practice.
   a. From 46 ft, the release point of Kim's pitches, estimate the time for Kim's pitches to cross the plate.
   b. What amount of time does it take for a 90-mph pitch to reach the plate from the standard 60 ft, 6 in.?
   c. How should the 50 ft, 6 in. distance for the batting-practice pitching rubber be changed so the time for a 55-mph pitch to cross the plate is the same as that for a 90-mph pitch?

*Corey Patterson taking batting practice.*

## TALE OF THE TAPE

Recall that I seem to be a cat on the prowl for applications of mathematics. Well I found yet another when I was attending a baseball game in Cincinnati. After a player had hit a home run into the stands over the 375-ft sign, an announcement ran across the scoreboard saying, "IBM Tale of the Tape—the ball would have traveled 415 ft had the ball not hit the stands." The subtle implication was that there was some kind of computerized method of making these calculations. As it turns out, the method was rather simple, involving some linear formulas.

How did I find out? Well, my co-author, Judy Beecher, asserts that it takes about ten phone calls to get to the root of an application and we are never afraid to make phone calls. I called IBM in Armonk, New York, and was quickly put in touch with John J. McMahon, then Manager of Corporate Promotions. He sent me a copy of a letter that had been sent to Roger Angell, noted baseball author, who was then associated with *New Yorker Magazine.* That letter and a subsequent conversation with its author, Robert K. Adair, gave me most of the answers to the question.

### The Mathematics behind the Tale of the Tape

In Figure 6, we see the trajectories of a well-hit baseball. Such a ball has backspin, which tends to make the trajectory look like a parabola, which is skewed away from home plate.

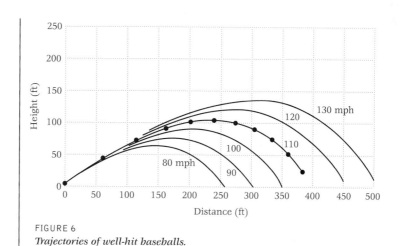

FIGURE 6
*Trajectories of well-hit baseballs.*

It should be emphasized that the trajectory, or flight, of a well-hit baseball (or golf ball) is *not* modeled by a parabola such as

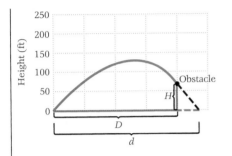

FIGURE 7
*Home run trajectories.*

$h(t) = -\frac{1}{2}gt^2 + (v_0 \sin\alpha)t + h_0$, a model often seen in calculus. In the case of this example, we will not develop the equations of the flight of a baseball, just consider Adair's results. Consider Figure 7. Note the flight of the ball. It hits an obstacle at a point $(D,H)$, where $D$ = the horizontal distance from home plate to where the ball is obstructed, and $H$ = the vertical distance from the ground to where the ball is obstructed. We want to estimate the distance, $d$, that the ball would have traveled had it not been obstructed. We do so with the following linear formula, which was developed by Adair.

### IBM Estimate of Distance

$d = kH + D$,

where

$D$ = horizontal distance, in feet, of the ball when obstructed,

$H$ = vertical height, in feet, of the ball when obstructed,

$d$ = estimated distance, in feet, ball would have traveled had it not been obstructed. Error: $\pm 10$ ft, and

$k$ = constant determined by the trajectory of ball.

We use three values of $k$, determined by the type of trajectory of the ball. See Figure 8. We make a judgment of the type of trajectory based on the launch angle of the ball when hit. This gives an estimate for $k$.

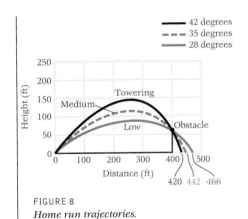

FIGURE 8
*Home run trajectories.*

### Distance Predictors

**Low trajectory (28°):**　　$d = 110\% \, H + D = 1.1H + D$,

**Medium trajectory (35°):**　$d = 70\% \, H + D = 0.7H + D$,

**High trajectory (42°):**　　$d = 50\% \, H + D = 0.5H + D$.

Suppose $D = 400$ ft and $H = 60$ ft. Then, using distance predictors, the estimated distances would be

**Low trajectory (28°):**　　$d = 110\% \, (60) + (400) = 466$ ft,

**Medium trajectory (35°):**　$d = 70\% \, (60) + (400) = 442$ ft,

**High trajectory (42°):**　　$d = 50\% \, (60) + (400) = 430$ ft.

Admittedly, deciding how to classify the trajectory of a home run ball requires an instant decision. Adair said, "If in doubt, take 70% of the height $H$ and add it to $D$." In this way, you can guess the Tale of the Tape from your seat. Once, during a Reds–Dodgers playoff, I did this after a home run and came within 1 ft of the answer. People around me couldn't believe it. You can do this as well! And, yes, this certainly was an .

**MPX**    MUSIC: "Willie, Mickey, and the Duke (Talking Baseball)" sung by Terry Cashman, Cashman, Lifesong 45086, 1981, audio diskette, *Greatest Baseball Hits*, Rhino, R4 70710.
MOVIE: *Field of Dreams* (1989), starring Kevin Costner and Amy Madigan. My second favorite baseball movie.

### Griffith Stadium Blast

There is an enduring grand story of a towering home run ball off the bat of Mickey Mantle in old Griffith Stadium in Washington, D.C. The ball hit a 40-ft high beer sign 460 ft from home plate. Our estimate would be that it would have traveled 480 feet. But the press blew the home run out of proportion, announcing that it would have traveled 565 feet. According to Adair, with a 20 to 40 mph wind, it could have traveled no farther than 511 feet.

*Mickey Mantle swings the bat.*

### Standard Conditions

Adair thinks that under standard conditions—70° temperatures, sea-level altitude, and no wind—450 ft is about the limit a normal man can hit a baseball. Of course a world-class weight lifter (say 6'8", 300 lb; picture Arnold Schwarzenegger) with catlike speed, taking two steps at the plate, using a 42", 56 oz bat, might be able to hit a batting-practice baseball more then 500 ft. In reality, though, if such a baseball player ever took two steps at the plate, the next pitch might be at his head because no pitcher would tolerate the insult of a hitter taking the extra step.

### The Rest of the Story

The question remained in my mind: What actually happens at the stadium when they announce the Tale of the Tape? I got the answer when I visited Dusty Baker after he became manager of the Giants. We were at Coors Field in Denver, home of the Colorado Rockies. I was given a press pass by the Giants. This allowed me to roam the ballpark at will, which I did with great joy. I ended up in the press box, where I watched players hitting homers. I decided to track down the source of announced "tales of the tape." What really happens is that the announcer is given a chart of the ballpark that lists the predicted lengths of home runs. He watches where the home runs go, checks the chart, and makes an announcement. To my chagrin, no computer enters the picture at the time of the home run. Oh well, such is the

*Mickey, Billy and the Cows.* I love baseball stories. One of my favorites was told by Mickey Mantle about his friend, Billy Martin. Billy was a firey ball player, and he brought his temperament with him into his managing days, getting into fights with his players or anybody who crossed him in a bar or elsewhere. He and Mickey Mantle were like brothers. One time they went hunting at a farm of a friend of Mickey's. Mick left Billy in the car and went inside to ask the farmer permission to hunt on his place.

The farmer said yes, but as a favor would Mick please shoot that mule out in the barnyard—the mule was old, he was attached to it, and did not have the heart to kill it himself.

Mickey, being the fun-lover and kidder he was, decided to play a trick on Billy. Mick went to the car and told Billy that the farmer would not let them hunt and that he was so mad, he was going to shoot the farmer's mule. So, off Mick went and shot the mule. At about the same time, he heard 3 other shots and went running back to Billy in concern. Billy said, "I shot 3 of his cows as well!!"

REFERENCES
Feller, B., and Gilbert, B., *Now Pitching: Bob Feller*. New York, Carol Publishing Group, 1990, pp. 120–122.

nature of marketing. This is truly an IRU application for me. I hope it is for you.

### Pursuing Further

1. As the photo shows, on May 22, 1963, Mickey Mantle hit a home run off the facade of the upper right-field deck of Yankee Stadium. It was a *towering* shot off Kansas City pitcher Bill Fischer. How far would it have traveled had it not hit the facade? Let $D = 367$ ft, though it is shown otherwise.

## THE LOST WAR YEARS

I met Bob Feller, Hall-of-Fame Pitcher for the Cleveland Indians, at a Los Angeles Dodgers Adult Fantasy Camp, being coached by all Hall-of-Fame players. (Incidentally, as a hitter I had to go to the plate when Bob Gibson was pitching. Needless to say, he struck me out.)

*Bob Feller, Cleveland Indians.*

I was a Cleveland Indians fan as a boy growing up in the early fifties when Feller was nearing the end of his career. I was thrilled to meet him at the camp. Prior to that, I had read Feller's book, *Now Pitching: Bob Feller*, in which he discusses the war years when he and several prominent players such as Ted Williams, Johnny Mize, Joe DiMaggio, and Hank Greenberg left the playing field to serve in World War II. Feller explained that an analyst in Seattle, Ralph Winnie, had analyzed his records and deduced that Feller would have had five no-hitters instead of three if the war had not interrupted his career. Winnie also projected that Feller would have had 373 career wins instead of 266.

Winnie also analyzed Ted Williams's case and projected that Ted would have had 2,663 runs batted in (RBI) instead of 1,839 (see Table 1), and would have been the all-time career leader ahead of Hank Aaron, who now holds the record at 2,297. Williams also would rank second to Aaron in career home runs with 743 to Aaron's 755.

Once again I set out on my own to agree or disagree with Winnie's results mathematically.

Using a TI-83+ graphing calculator, I analyzed the data in Table 1 by fitting the data to a linear function by regression. The function is given by

$$y = -0.402x + 813.785$$

where $x$ = the year, and $y$ = the predicted number of home runs in year $x$. Then the function was used to estimate the home runs in the war years of 1943, 1944, and 1945 (World War II), and 1952 and 1953 (Korean War). I deleted the HRs he got, 14, for the predicted ones. The total of these predicted HRs was 662.

Interestingly, if we just take the average for the complete years Williams played, we get 34 home runs per year.

$$507/15 = 34, \qquad 34 \times 5 = 170,$$
$$521 - 14 + 34 \times 5 = 677$$

Multiplying by 5, subtracting 14, and adding to 521, we get a third estimate, 677, which was close to the estimate found using the preceding linear function. Comparing all the estimates in the last line of Table 1, I could not corroborate the results Winnie predicted for Williams, which was 743 home runs.

No other reference was given to Ralph Winnie in Feller's book other than that he lived in Seattle, so I decided to contact Feller. I wrote a letter to his publisher and a letter without a street address to his home city of Gates Mills, Ohio.

*Ted Williams, Boston Red Sox, Best Hitter of All Time.*

**Table 1**    Ted Williams's Home Runs

| Year | Actual Home Runs | Home Runs Predicted by $y = -0.402x + 813.785$ | Home Runs Predicted by Averaging |
|------|------|------|------|
| 1939 | 31 | | |
| 1940 | 23 | | |
| 1941 | 37 | | |
| 1942 | 36 | | |
| 1943 | 0 (WWII) | (33) | (32) |
| 1944 | 0 (WWII) | (32) | (32) |
| 1945 | 0 (WWII) | (32) | (32) |
| 1946 | 38 | | |
| 1947 | 32 | | |
| 1948 | 25 | | |
| 1949 | 43 | | |
| 1950 | 28 (hurt) | | |
| 1951 | 30 | | |
| 1952 | 1 (Korean War) | (29) | (32) |
| 1953 | 13 (Korean War) | (29) | (32) |
| 1954 | 29 | | |
| 1955 | 28 | | |
| 1956 | 24 | | |
| 1957 | 38 | | |
| 1958 | 26 | | |
| 1959 | 10 | | |
| 1960 | 29 | | |
| Totals | 521 | 662 | 677 |
| | | $662 < 743$ | $507 / 15 = 34$,  $34 \times 5 = 170$, |
| | | | $521 - 14 + 34 \times 5 = 677$, |
| | | | $677 < 743$ |

*A man has to have goals—for a day, for a lifetime—and that was mine, to have people say, 'There goes the greatest hitter who ever lived.'*
—TED WILLIAMS

*There's only one way to become a hitter. Go up to the plate and get mad. Get mad at yourself and mad at the pitcher.*
—TED WILLIAMS

I called Feller and asked about Ralph Winnie. Feller told me Winnie was in an organization known as SABR, Society for American Baseball Research, an organization for baseball enthusiasts who analyze the game to its absolute literary-historical-statistical depths. Its publications include articles about such diverse topics as the history of the Negro leagues and the effects of the DH (designated hitter) on the game. I knew about SABR because I was a member. He gave me Winnie's phone number.

*I must tell a related, amusing story. My beloved wife, Elaine, has never had an interest in baseball and knows very few famous players. One Saturday, after returning from running errands, I asked as usual if there had been any calls. She said in a very casual manner, "Bob Feller called." I reacted in amazement, "Bob Feller? Bob Feller called me? Wow!" To me, it was like I had just gotten a call from the President. To Elaine, it was like a call from a friend down the street.*

I called Winnie and informed him that my mathematical results seemed to refute his predictions. Winnie was irritated with me, telling me that I had not come anywhere close to the careful analysis he had carried out, but he sent me his book. Without offering details of the mathematics used, Winnie asserts that he considered many factors such as the effects of playing 162 games versus 154 games per season in the era of Feller and Williams. He also considered odd facts such as the practice of many teams keeping top-notch players hidden in the minor leagues and the effects of catastrophic problems that shortened careers of excellent players like Lou Gehrig (ALS disease) and Herb Score (hit in the eye by a shot from the bat of Gil McDougald). If Winnie were to repeat his study today, it might include such new features as the effect of expansion teams and the use of the designated hitter.

According to Winnie's calculations, Table 2 shows the would-be home run totals of the top four home run hitters.

REFERENCES
Winnie, R., *What If?* published by Ralph Winnie, 17905 3rd Ave NW, Seattle, WA 98177, 1986.

**Table 2**

| Player | Actual Career Home Runs | Projected Career Home Runs |
|---|---|---|
| Ted Williams | 521-10th | 780 |
| Hank Aaron | 755-1st | 768 |
| Babe Ruth | 714-2nd | 752 |
| Lou Gehrig | 493 | 742 |
| Frank Robinson | 586-4th | 670 |

*Source: What If?, by Ralph Winnie, p. 46.*

I also did my analysis of the career RBI totals of Ted Williams. Although my stats were again short of Winnie's (Table 4), I did corroborate that Williams would hold the all-time RBI record over Aaron. My results are in Table 3.

In his book, Bob Feller noted that he had not one regret about the playing time or the statistics he lost serving in World War II. He was proud to have served his country in the U.S. Navy. Even the damage he suffered to his feet and legs standing on the hard decks of ships did not deter his patriotism.

The projections are, in truth, a great application to the baseball-math enthusiast, but only God knows what might have happened without interruptions for military service.

**Table 3**    Ted Williams's RBIs

| Year | Actual RBIs | RBIs Predicted by $y = -3.354x + 6,646.484$ | RBIs Predicted by Averaging |
|------|------|------|------|
| 1939 | 145 | | |
| 1940 | 113 | | |
| 1941 | 120 | | |
| 1942 | 137 | | |
| 1943 | 0 (WWII) | (130) | (108) |
| 1944 | 0 (WWII) | (127) | (108) |
| 1945 | 0 (WWII) | (123) | (108) |
| 1946 | 123 | | |
| 1947 | 114 | | |
| 1948 | 127 | | |
| 1949 | 159 | | |
| 1950 | 97 (hurt) | | |
| 1951 | 126 | | |
| 1952 | 3 (Korean War) | (100) | (108) |
| 1953 | 34 (Korean War) | (96) | (108) |
| 1954 | 89 | | |
| 1955 | 83 | | |
| 1956 | 82 | | |
| 1957 | 87 | | |
| 1958 | 85 | | |
| 1959 | 43 | | |
| 1960 | 72 | | |
| Totals | 1839 | 2378 | 2402 |
| | | 2378 > 2297 | 1802/15 = 120,  120 × 5 = 600, 1839 − 37 + 120 × 5 = 2402 |

**Table 4**    Actual vs. Projected RBIs

| Player | Actual | Projected Career RBIs |
|------|------|------|
| Ted Williams | 1839 | 2795 |
| Lou Gehrig | 1990-3rd | 2659 |
| Joe DiMaggio | 1537 | 2357 |
| Hank Aaron | 2297-1st | 2336 |
| Babe Ruth | 2209-2nd | 2327 |

*Source:* What If?, *by Ralph Winnie, pp. 65–66.*

***Pursuing Further***

1. *Lou Gehrig.* Use the following data on Lou Gehrig's career and a graphing calculator for the years 1925–1938, excluding 1939 when he was ill. Fit a linear regression function to the data and predict Gehrig's home runs for years 1939–1942, and his career home run total.

Lou Gehrig, first base

## Hitting Stats

| SEASON | TEAM | G | AB | R | H | 2B | 3B | HR | RBI | TB | BB | SO | SB | CS |
|---|---|---|---|---|---|---|---|---|---|---|---|---|---|---|
| 1923 | New York Yankees | 13 | 26 | 6 | 11 | 4 | 1 | 1 | 9 | 20 | 2 | 5 | 0 | 0 |
| 1924 | New York Yankees | 10 | 12 | 2 | 6 | 1 | 0 | 0 | 5 | 7 | 1 | 3 | 0 | 0 |
| 1925 | New York Yankees | 126 | 437 | 73 | 129 | 23 | 10 | 20 | 68 | 232 | 46 | 49 | 6 | 3 |
| 1926 | New York Yankees | 155 | 572 | 135 | 179 | 47 | 20 | 16 | 112 | 314 | 105 | 73 | 6 | 5 |
| 1927 | New York Yankees | 155 | 584 | 149 | 218 | 52 | 18 | 47 | 175 | 447 | 109 | 84 | 10 | 8 |
| 1928 | New York Yankees | 154 | 562 | 139 | 210 | 47 | 13 | 27 | 142 | 364 | 95 | 69 | 4 | 11 |
| 1929 | New York Yankees | 154 | 553 | 127 | 166 | 32 | 10 | 35 | 126 | 323 | 122 | 68 | 4 | 4 |
| 1930 | New York Yankees | 154 | 581 | 143 | 220 | 42 | 17 | 41 | 174 | 419 | 101 | 63 | 12 | 14 |
| 1931 | New York Yankees | 155 | 619 | 163 | 211 | 31 | 15 | 46 | 184 | 410 | 117 | 56 | 17 | 12 |
| 1932 | New York Yankees | 156 | 596 | 138 | 208 | 42 | 9 | 34 | 151 | 370 | 108 | 38 | 4 | 11 |
| 1933 | New York Yankees | 152 | 593 | 138 | 198 | 41 | 12 | 32 | 139 | 359 | 92 | 42 | 9 | 13 |
| 1934 | New York Yankees | 154 | 579 | 128 | 210 | 40 | 6 | 49 | 165 | 409 | 109 | 31 | 9 | 5 |
| 1935 | New York Yankees | 149 | 535 | 125 | 176 | 26 | 10 | 30 | 119 | 312 | 132 | 38 | 8 | 7 |
| 1936 | New York Yankees | 155 | 579 | 167 | 205 | 37 | 7 | 49 | 152 | 403 | 130 | 46 | 3 | 4 |
| 1937 | New York Yankees | 157 | 569 | 138 | 200 | 37 | 9 | 37 | 159 | 366 | 127 | 49 | 4 | 3 |
| 1938 | New York Yankees | 157 | 576 | 115 | 170 | 32 | 6 | 29 | 114 | 301 | 107 | 75 | 6 | 1 |
| 1939 | New York Yankees | 8 | 28 | 2 | 4 | 0 | 0 | 0 | 1 | 4 | 5 | 1 | 0 | 0 |
| Career Totals | | 2164 | 8001 | 1888 | 2721 | 534 | 163 | 493 | 1995 | 5060 | 1508 | 790 | 102 | 101 |

*Source: Major League Baseball; www.mlb.com*

2. *J. R. Richard.* An imposing 6'8" figure on the mound, J. R. Richard also had a fastball clocked up to 100 mph, but his career ended on July 30, 1980, when he suffered a stroke. Dusty Baker said he was the pitcher the players feared most in his era. The following table shows his career stats from 1971 to 1980. Use the data for Richard's wins in the years 1975–1979 and a graphing calculator to fit a linear regression function to the data. Then predict his wins for the years 1980–1991 and compute his career win total.

J.R. Richard, Pitcher

**Pitching Stats**

| SEASON | Team | W | L | ERA | G | GS | CG | SHO | SV | SVO | IP | H | R | ER | HR |
|--------|------|---|---|-----|---|----|----|----|----|-----|----|---|---|----|----|
| 1971 | Houston Astros | 2 | 1 | 3.43 | 4 | 4 | 1 | 0 | 0 | — | 21.0 | 17 | 9 | 8 | 1 |
| 1972 | Houston Astros | 1 | 0 | 13.50 | 4 | 1 | 0 | 0 | 0 | — | 6.0 | 10 | 9 | 9 | 0 |
| 1973 | Houston Astros | 6 | 2 | 4.00 | 16 | 10 | 2 | 1 | 0 | — | 72.0 | 54 | 37 | 32 | 2 |
| 1974 | Houston Astros | 2 | 3 | 4.18 | 15 | 9 | 0 | 0 | 0 | — | 64.2 | 58 | 31 | 30 | 3 |
| 1975 | Houston Astros | 12 | 10 | 4.39 | 33 | 31 | 7 | 1 | 0 | — | 203.0 | 178 | 107 | 99 | 8 |
| 1976 | Houston Astros | 20 | 15 | 2.75 | 39 | 39 | 14 | 3 | 0 | — | 291.0 | 221 | 105 | 89 | 14 |
| 1977 | Houston Astros | 18 | 12 | 2.97 | 36 | 36 | 13 | 3 | 0 | — | 267.0 | 212 | 94 | 88 | 18 |
| 1978 | Houston Astros | 18 | 11 | 3.11 | 36 | 36 | 16 | 3 | 0 | — | 275.1 | 192 | 104 | 95 | 12 |
| 1979 | Houston Astros | 18 | 13 | 2.71 | 38 | 38 | 19 | 4 | 0 | — | 292.1 | 220 | 96 | 88 | 13 |
| 1980 | Houston Astros | 10 | 4 | 1.90 | 17 | 17 | 4 | 4 | 0 | — | 113.2 | 65 | 31 | 24 | 2 |
| Career Totals | | 107 | 71 | 3.15 | 238 | 221 | 76 | 19 | 0 | — | 1606.0 | 1227 | 625 | 562 | 73 |

*Source: Major League Baseball*

## BASEBALL STATISTICS

Since baseball is the sport in which statistics are most analyzed, I am going to relate them to my experiences with Dusty Baker and the San Francisco Giants.

A player's *batting average,* BA or AVG, is defined as follows:

$$BA = AVG = \frac{\text{Total Number of Hits}}{\text{Official at Bats}} = \frac{H}{AB}$$

where walks, sacrifice-fly RBIs, sacrifice bunts, times being hit by a pitch, are not official at bats (AB). Dusty Baker refers to the latter as *non-at-bats.*

For example, a player who gets 3 hits in 10 official at bats has a 0.300 batting average. Note that he has actually failed in 7 out of 10 attempts. Where else can you fail that often and make a salary of several million dollars playing a kid's game? Any hitter with a 0.300 batting average or better is considered an excellent hitter. An average of say 0.270 to 0.299 is a good hitter, and below 0.270, a marginal or weak hitter. Table 5 shows the batting averages of the top seven players in the major leagues in 2003.

**Table 5**  MLB Leaders: Batting Average, 2003 Regular Season

| Rank | Name | AVG | AB | H |
|------|------|-----|-----|-----|
| 1 | Albert Pujols, StL | .359 | 591 | 212 |
| 2 | Todd Helton, Col | .358 | 583 | 209 |
| 3 | Barry Bonds, SF | .341 | 390 | 133 |
| 4 | Edgar Renteria, StL | .330 | 587 | 194 |
| 5 | Gary Sheffield, Atl | .330 | 576 | 190 |
| 6 | Bill Mueller, Bos | .326 | 524 | 171 |
| 7 | Manny Ramirez, Bos | .325 | 569 | 185 |

*AVG = batting average, AB = at bat, H = hits.*

Because of the small numbers in H/AB, there can be great variation in a player's average at the beginning of the season, but very little at the end of the season when the numbers are very large, in the hundreds. For example,

$$\text{Effect of 1 hit early:} \quad \frac{3}{8} = 0.375 \quad \Rightarrow \quad \frac{3+1}{8+1} = \frac{4}{9} \approx 0.444$$

$$\text{Effect of 1 hit late:} \quad \frac{189}{504} = 0.375 \quad \Rightarrow \quad \frac{180+1}{504+1} = \frac{190}{505} \approx 0.376$$

Figure 9 is a graph of what I consider a typical batting average over a season. Suppose a player's batting average for the season ends up at 0.280. The wavy blue line represents the changes in the true batting average of a player. The solid blue curve represents a continuously decreasing function which might be used to "model" the player's batting average over the season in relation to his final average.

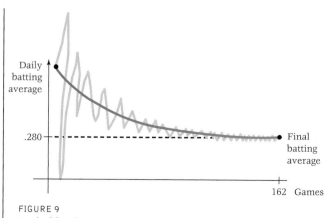

FIGURE 9
*Typical batting average for 1 season.*

This model has been drawn from my experience following baseball over 50 years. No extensive data analysis has been done, though the model is somewhat corroborated by the following graphs of BAs for two players. Remember, some players start with a low average and increase it over a season.

## Luis Gonzalez
#20 | <u>Left Field</u> | <u>Arizona Diamondbacks</u>

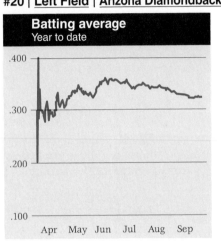

## Ichiro Suzuki
#51 | <u>Right Field</u> | <u>Seattle Mariners</u>

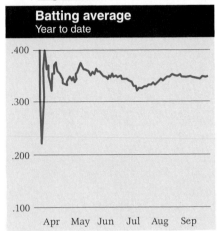

**FIGURE 10**
*Batting averages of Luis Gonzalez and Ichiro Suzuki in a recent year.*

The solid blue curve in Figure 9 is not a perfect model and is conceived out of my intuition, not a careful statistical analysis. Nevertheless, you will see Dusty Baker and his players use this model. What happens is that a player's BA starts out, in this case, quite high and wavers rapidly early in the season. Over the course of the season, as shown by the blue curve, the BA tends to *decrease* to the final BA. If you know statistics, you will recognize the Law of Large Numbers in this idea, but we will not discuss that here.

Dusty Baker was aware of this as a player, and in his capacity now as a manager, he makes his players aware of the curve. In effect, players are taught to realize that when you get a non-at-bat, you tend to keep your batting average higher, because BA tends to be a continuously decreasing function for players with early high averages. So if you make any kind of non-at-bat, draw a walk, make a sacrifice bunt, and so on, you are keeping your average higher. Quite simply, you have not received a regular at bat, which tends to decrease your average. Compare these two situations:

3 hits in 9 at bats in 3 games gives a 0.333 BA.

3 hits in 12 at bats in those three games gives a 0.250 BA.

Suppose the hitter had 3 non-at-bats out of those 12 opportunities. Then the hitter would look a lot better at 0.333 than at 0.250.

*Pete Rose*

REFERENCE

Will, G., *Men at Work,* New York, Macmillan Publishing Company, 1990. pp. 316–317.

Pete Rose is probably my favorite all-time player on the field. Off the field his character has left something to be desired. (And, yes, I do think he should be forgiven for his gambling allegations and voted into the Hall of Fame). Dusty said, and I agree from all the Reds games I followed during Rose's career, that Pete Rose was probably the "greediest" hitter to ever play the game. If he got 1 hit in his first at bat, you could almost count on him going 2/3, 3/4, or 4/4 for the game. He always went to the plate eager, intense, and aggressive no matter how many previous hits he had made or the score of the game.

According to Dusty, immature players often fall into a trap. Say a player gets 2 hits in his first 2 at bats and ends up 2/4 for the night; he has a very respectable .500 for the game. Young players in those situations often do what I call "hit and giggle" for the rest of the game. They know after those first two at bats that they will have a good night, so they don't bear down on the next at bats. But suppose that hitter faces a tough pitcher the next night. Then

$$\left.\begin{array}{c} 2/4 \\ 0/4 \end{array}\right\} \Rightarrow 2/8 = 0.250$$

for the two nights—not so good! If that player had played greedy like Pete Rose, then maybe the situation would change to

$$\left.\begin{array}{c} 3/4 \\ 0/4 \end{array}\right\} \Rightarrow 3/8 = 0.375$$

for the two nights—looking good!

The moral of the story from Dusty is, no matter what your playing level, *never never give away an at bat!*

## Standard Deviations and the Demise of the 0.400 Hitter

George Will, in his book *Men at Work,* referred to studies by the late Stephen Jay Gould, eminent Harvard scientist and baseball fan, on evolutionary processes in batting.

$$\text{BA in 1870s} = \mu = 0.260$$

$$\text{BA in 1989} = \mu = 0.255$$

According to Gould, the batting averages of players have essentially not decreased by much over the years. But the variation, or standard deviation, of the batting averages has decreased considerably, as modeled in Figure 11.

This is a fascinating result to those who know the mathematics, but unfortunately it is difficult to explain standard deviation to the average person on the street who has no training in the concept. Certainly, no announcer comments about changes in a hitter's stan-

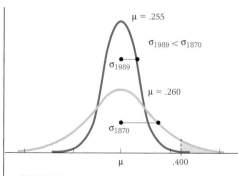

FIGURE 11
*Means and standard deviations of batting averages.*

*Jay Johnstone, Jerry Reuss, and Tommy LaSorda.*
*Jay Johnstone and Jerry Reuss played for Tommy*
*LaSorda when he was manager of the Dodgers.*
*Johnstone and Reuss were born pranksters. Once*
*the two of them borrowed clothes from the Dodgers'*
*field maintenance crew and went out and raked the*
*infield between innings. LaSorda was livid when he*
*discovered the deed.*

*Johnstone was once my manager at Dodger*
*Baseball camp, but we weren't hitting very well. He*
*actually took all our baseball bats, stacked them in a*
*pile, poured gasoline on them, and lit them in order to*
*"fire up our hitting." He got our attention, but our*
*hitting continued in futility.*

*Johnstone was a prankster extraordinaire. One*
*prank he tells at Dodger Camp involved Tommy*
*LaSorda and his enjoyment of food.*

*Tommy never missed a meal. One night during*
*spring training, Johnstone and Reuss snuck into his*
*room and removed the voice-transmitting device from*
*the telephone. Thus, Tommy could hear the operator*
*but she could not hear him.*

*After Tommy went to bed that night, Johnstone*
*and Reuss got some rope, tied one end to the door knob*
*and the other end to a tree; there was no way Tommy*
*could get out of his room the next morning. When*
*Tommy failed to get the door open, he tried the*
*windows, but they were too small to crawl through.*
*Furious, he called the front desk and yelled into the*
*phone but the operator could not hear him. After some*
*time, someone saw the rope and let him out. He not*
*only missed his breakfast, but he was late to the team*
*bus, a major no-no he impressed on his players.*

dard deviation. When explaining standard deviation to someone, I like to use the notion of a "spread" number and try to tell how numbers are spread out on both sides of the mean, or average.

Think of the probability of a hitter batting 0.400 or better as the area under each curve to the right of 0.400 on the *x*-axis in Figure 11. Note that in the 1870s the spread number was quite large and the shaded area to the right of 0.400 under the curve for the 1870 hitters is much higher than the blue to the right of 0.400 under the curve for the 1989 hitters. These areas are probabilities. Thus, there was a much smaller chance of a hitter batting 0.400 in 1989 than in 1870. Gould asserts that the reasons for the demise of the 0.400 hitter is that the game has refined itself, the equipment is better, the fields are better, and the instruction is better. But to me, these assertions would apply to pitching as well as hitting, so I can give no sound reason other than to accept the results of Gould's studies.

Look back again at the solid blue curve in Figure 9. I find it amusing, if not irritating, that if a player has a 0.400 batting average or better in June or July, then the media go crazy analyzing and discussing the possibility. Since that blue curve is a decreasing function, the media shouldn't involve themselves in such foolish distractions unless the player is hitting over 0.400 by the first of September. This overreaction occurred with George Brett in 1980 with the Kansas City Royals; Brett finished with a 0.390 BA. The overreaction occurred again with Tony Gwynn of the San Diego Padres in 1994; Gwynn finished with a 0.394 BA.

## Slugging Percentage

A player's *slugging percentage* (SLG), or *slugging average,* is defined as

$$\text{Slugging Percentage} = \text{SLG} = \frac{\text{Total Bases}}{\text{Official at Bats}}$$

$$= \frac{4(HR) + 3(T) + 2(D) + S}{AB},$$

where for the season,

$HR$ = the total number of home runs,

$T$ = the total number of triples,

*Barry Bonds*

$D$ = the total number of doubles and

$S$ = the total number of singles.

Barry Bonds, of the San Francisco Giants, had probably the most incredible season in the history of baseball in 2001, highlighted by his record-breaking 73 home runs, eclipsing the all-time season total record of 70 by Mark McGwire. He also broke the all-time record for slugging average in a season. Let's compute his SLG. His hitting statistics are in Table 6.

**Table 6**  Career Batting Statistics of Barry Bonds

| Year | Team | Avg | G | AB | R | H | 2B | 3B | HR | RBI | BB | K | OBP | SLG | OPS |
|------|------|-----|---|----|---|---|----|----|----|-----|----|---|-----|-----|-----|
| 1986 | PIT | .223 | 113 | 413 | 72 | 92 | 26 | 3 | 16 | 48 | 65 | 102 | .330 | .416 | .746 |
| 1987 | PIT | .261 | 150 | 551 | 99 | 144 | 34 | 9 | 25 | 59 | 54 | 88 | .329 | .492 | .821 |
| 1988 | PIT | .283 | 144 | 538 | 97 | 152 | 30 | 5 | 24 | 58 | 72 | 82 | .368 | .491 | .859 |
| 1989 | PIT | .248 | 159 | 580 | 96 | 144 | 34 | 6 | 19 | 58 | 93 | 93 | .351 | .426 | .777 |
| 1990 | PIT | .301 | 151 | 519 | 104 | 156 | 32 | 3 | 33 | 114 | 93 | 83 | .406 | .565 | .970 |
| 1991 | PIT | .292 | 153 | 510 | 95 | 149 | 28 | 5 | 25 | 116 | 107 | 73 | .410 | .514 | .924 |
| 1992 | PIT | .311 | 140 | 473 | 109 | 147 | 36 | 5 | 34 | 103 | 127 | 69 | .456 | .624 | 1.080 |
| 1993 | SF | .336 | 159 | 539 | 129 | 181 | 38 | 4 | 46 | 123 | 126 | 79 | .458 | .677 | 1.136 |
| 1994 | SF | .312 | 112 | 391 | 89 | 122 | 18 | 1 | 37 | 81 | 74 | 43 | .426 | .647 | 1.073 |
| 1995 | SF | .294 | 144 | 506 | 109 | 149 | 30 | 7 | 33 | 104 | 120 | 83 | .431 | .577 | 1.009 |
| 1996 | SF | .308 | 158 | 517 | 122 | 159 | 27 | 3 | 42 | 129 | 151 | 76 | .461 | .615 | 1.076 |
| 1997 | SF | .291 | 159 | 532 | 123 | 155 | 26 | 5 | 40 | 101 | 145 | 87 | .446 | .585 | 1.031 |
| 1998 | SF | .303 | 156 | 552 | 120 | 167 | 44 | 7 | 37 | 122 | 130 | 92 | .438 | .609 | 1.047 |
| 1999 | SF | .262 | 102 | 355 | 91 | 93 | 20 | 2 | 34 | 83 | 73 | 62 | .389 | .617 | 1.006 |
| 2000 | SF | .306 | 143 | 480 | 129 | 147 | 28 | 4 | 49 | 106 | 117 | 77 | .440 | .688 | 1.127 |
| 2001 | SF | .328 | 153 | 476 | 129 | 156 | 32 | 2 | 73 | 137 | 177 | 93 | .515 | .863 | 1.378 |
| 2002 | SF | .370 | 143 | 403 | 117 | 149 | 31 | 2 | 46 | 110 | 198 | 47 | .582 | .799 | 1.381 |
| 2003 | SF | .341 | 130 | 390 | 111 | 133 | 22 | 1 | 45 | 90 | 148 | 58 | .529 | .749 | 1.278 |
| Totals | | .297 | 2569 | 8725 | 1941 | 2595 | 536 | 74 | 658 | 1742 | 2070 | 1387 | .433 | .602 | 1.035 |

From Table 6 we see that $HR$ = 73, $T$ = 2, $D$ = 32, and $S$ = 49. Table 6 does not list the actual number of singles. We must take total hits, 156, and subtract home runs, triples, and doubles, to get the singles. Then

$$\text{SLG} = \frac{4(HR) + 3(T) + 2(D) + S}{AB}$$
$$= \frac{4(73) + 3(2) + 2(32) + 49}{476}$$
$$\approx 0.863.$$

*Bruce Froemming, Pete Rose, and the Birds.* *Another of my favorite baseball stories is told often by major-league umpire Bruce Froemming at the Dodger Fantasy Camp. Many people consider Froemming to be the finest ball-strike umpire in the game.*

*Bruce came to the major leagues in 1970 and served his first year with Al Barlick, the noted umpire, who was completing his last year in the majors. It is typical of players and umpires to play pranks on rookies. Earlier that season, Nate Colbert of the San Diego Padres had had the misfortune to have a fly ball hit a bird in the outfield on its way to what might have been a home run. No ground rule covered this play. It caused quite a controversy.*

*To pull a prank on rookie Froemming, Barlick informed him that for a couple of weeks he should go out on the diamond one hour before each game and "look for birds." Wanting to impress, Froemming did as instructed with diligence, until finally unraveling the prank. I imagine him out there staring at the sky.*

*Not long after this, the umpire team went to Cincinnati for a series between the Reds and the Mets. Pete Rose, the captain of the Reds at the time, brought out his lineup card to meet with the umpires and the opposing umpire, Gil Hodges. This time the prank was on Rose and was committed by Hodges and the other umpires.*

*To put Froemming on the other side of the joke, the umpires told him to tell Rose something like the following, "Pete, because of the bird incident in San Diego, I have been assigned to watch for birds in the ballparks. I noticed here that there are two pigeons flying around in right field so we need a ground rule to cover the situation." Pete agreed and asked what the rule would be. Froemming says, "There are two birds, a male and a female. If you hit the male, it is a ground rule double. If you hit the female, it is all you can get."*

A hitter is considered to have an excellent slugging percentage if it is 0.600 or better. So, Bonds set a phenomenal record.

## OBP and OPS Statistics

Baseball statisticians also consider a statistic called *on-base percentage*, OBP. It is an adaptation of BA that takes into account other ways of getting on base such as walking and being hit by a pitch. OBP is defined as follows.

$$OBP = \frac{H + W + HBP}{AB + W + HBP + SF},$$

where

$H$ = total number of hits,

$W$ = total number of walks (often abbreviated BB, for number of bases on balls),

$HBP$ = total number of times batter was hit by a pitch,

$SF$ = total number of sacrifice flys, and

$AB$ = total at bats.

Most books and Internet references do not list HBP.

Ted Williams extolled the use of the statistic *production* OPS, defined simply as on-base percentage plus slugging percentage. That is,

$$OPS = OBP + SLG$$
$$= \frac{H + W + HBP}{AB + W + HBP + SF} + \frac{4(HR) + 3(T) + 2(D) + S}{AB},$$

where for the season,

$HR$ = the total number of home runs,

$T$ = the total number of triples,

$D$ = the total number of doubles,

$S$ = the total number of singles,

$H$ = total number of hits,

$W$ = total number of walks (often abbreviated BB, for number of bases on balls),

$HBP$ = total number of times batter was hit by a pitch,

$SF$ = total number of sacrifice flys, and

$AB$ = total at bats.

Williams asserted adamantly, "I realize that everyone has a different idea of what constitutes a great hitter. For some it's a high batting average. For others, it's the guy with the most total hits, or home runs, or RBIs. I've al-

REFERENCE
Williams, T., and Prime, J., *Ted Williams' Hit List*, Indianapolis, Indiana, Masters Press, A Division of Howard W. Sams & Co., 1996, p. 33.

ways believed that slugging percentage plus on-base percentage is absolutely the best way to rate the hitters."

To me Ted Williams is "absolutely" the greatest hitter to ever play the game. If he says OPS is the best stat to rate a hitter, I cannot disagree. In most books and Internet sources both OBP and SLG are given, so we can simply add to find OPS.

For Barry Bonds's outstanding 2001 season, his OBP was 0.515 and his SLG was 0.863. Thus, his OPS was 0.515 + 0.863, or 1.378. Williams asserts that any player with an OPS above 1.0 has an excellent season. In their careers, very few players had an OPS above 1; Babe Ruth had 1.163, Lou Gehrig had 1.080, Mickey Mantle had 0.979, and Willie Mays had 0.944. Table 7 shows the OPS statistics for the top five players in the major leagues in 2001.

**Table 7**     MLB Leaders: On-Base plus Slugging Percentages, (OPS), 2001 Regular Season

| Rank | Name | Team | OPS | AVG | RBI | AB | HR | OBP | SLG |
|---|---|---|---|---|---|---|---|---|---|
| 1 | Barry Bonds | SF | 1.379 | .328 | 137 | 476 | 156 | .515 | .863 |
| 2 | Sammy Sosa | ChC | 1.174 | .328 | 160 | 577 | 189 | .437 | .737 |
| 3 | Jason Giambi | NYY | 1.137 | .342 | 120 | 520 | 178 | .477 | .660 |
| 4 | Luis Gonzalez | Ari | 1.117 | .325 | 142 | 609 | 198 | .429 | .688 |
| 5 | Todd Helton | Col | 1.116 | .336 | 146 | 587 | 197 | .434 | .685 |

## Pitcher's Earned Run Average

A pitcher's win–loss record is the first statistic used to evaluate the player. The second most used statistic is the earned run average, ERA. In 2001 the outstanding pitcher for the Arizona Diamondbacks, Randy Johnson, gave up 69 earned runs in $249\frac{2}{3}$ innings of pitching. The mixed numeral $249\frac{2}{3}$ means that he pitched 249 full innings and plus 2 outs of 3 possible outs. The ERA is based on how many earned runs the pitcher gives up every 9 innings. To find the ERA, we set up a proportion:

$$\frac{\text{ERA}}{9} = \frac{69}{249\frac{2}{3}}.$$

Then we solve the proportion by multiplying both sides by 9, and carry out the calculations:

$$\frac{\text{ERA}}{9} = \frac{69}{249\frac{2}{3}}$$

$$\text{ERA} = 9 \cdot \frac{69}{249\frac{2}{3}}$$

$$= 2.49 \qquad \text{Rounded to the nearest hundredth}$$

This leads us to the definition of a pitcher's earned run average, ERA:

$$\text{ERA} = 9 \cdot \frac{ER}{IP},$$

where

ER = the number of earned runs allowed,

IP = the number of innings pitched

A pitcher with an ERA below 3.00 is considered excellent, although in the present era of weaker pitching an ERA of 4.00 or less is considered good. Table 8 shows the ERA statistics for the top five pitchers in the major leagues in 2001.

**Table 8**   MLB Leaders ERA, 2001 Regular Season

| Player | TEAM | W | L | ERA | IP | H | R | ER | HR |
|--------|------|---|---|-----|-----|-----|-----|-----|-----|
| 1. R. Johnson | ARI | 21 | 6 | 2.49 | 249.2 | 181 | 74 | 69 | 19 |
| 2. C. Schilling | ARI | 22 | 6 | 2.98 | 256.2 | 237 | 86 | 85 | 37 |
| 3. J. Burkett | ATL | 12 | 12 | 3.04 | 219.1 | 187 | 83 | 74 | 17 |
| 4. F. Garcia | SEA | 18 | 6 | 3.05 | 238.2 | 199 | 88 | 81 | 16 |
| 5. G. Maddux | ATL | 17 | 11 | 3.05 | 233.0 | 220 | 86 | 79 | 20 |

*Note: When reading a table for pitchers, 249.2 IP means $249\frac{2}{3}$ and 219.1 IP means $219\frac{1}{3}$.*

### Pursuing Further

1. Below are the career totals of five all-time leading players in the major leagues through June 23, 2003. In each case compute their batting average, AVG, on-base percentage, OBP, slugging percentage, SLG, and their OPS.

| Player | AB | H | 2B | 3B | HR | W | SF | HBP | AVG | OBP | SLG | OPS |
|--------|-----|------|-----|-----|-----|------|------|-----|-----|-----|-----|-----|
| Hank Aaron | 12,364 | 3771 | 624 | 98 | 755 | 1402 | 121 | 32 | | | | |
| Babe Ruth | 8399 | 2873 | 506 | 136 | 714 | 2062 | N/A | 43 | | | | |
| Willie Mays | 10,881 | 3283 | 523 | 140 | 660 | 1464 | 91 | 44 | | | | |
| Barry Bonds | 8725 | 2595 | 536 | 74 | 658 | 2070 | 84 | 84 | | | | |
| Mark McGwire | 6187 | 1626 | 252 | 6 | 583 | 1317 | 78 | 75 | | | | |

N/A = Not available; use 0.

**2.** What is the highest that a slugging percentage can be? the lowest?

**3.** Below are the career totals of five all-time leading pitchers in the major leagues. Compute each ERA.

| PLAYER | YRS | G | IP | H | ER | W | L | ERA |
|---|---|---|---|---|---|---|---|---|
| 1. Ed Walsh | 14 | 430 | 2964.1 | 2346 | 598 | 195 | 126 | |
| 2. Addie Joss | 9 | 286 | 2327.0 | 1888 | 488 | 160 | 97 | |
| 3. Mordecai Brown | 14 | 481 | 3172.1 | 2708 | 725 | 239 | 130 | |
| 4. John Ward | 7 | 292 | 2461.2 | 2317 | 575 | 164 | 102 | |
| 5. Christy Mathewson | 17 | 635 | 4780.2 | 4218 | 1133 | 373 | 188 | |

**4.** Below are five all-time single-season ERA record-holding pitchers. Compute each ERA.

| PLAYER | YRS | G | IP | H | ER | W | L | ERA |
|---|---|---|---|---|---|---|---|---|
| 1. Dutch Leonard, Bos | 1914 | 36 | 224.2 | 139 | 24 | 19 | 5 | |
| 2. Mordecai Brown, Chi | 1906 | 36 | 277.1 | 198 | 32 | 26 | 6 | |
| 3. Bob Gibson, StL | 1968 | 34 | 304.2 | 198 | 38 | 22 | 9 | |
| 4. Christy Mathewson, NY | 1909 | 37 | 275.1 | 192 | 35 | 25 | 6 | |
| 5. Walter Johnson, Was | 1913 | 48 | 346.0 | 232 | 44 | 36 | 7 | |

| Innings Pitched (i) | Earned Run Average (E) |
|---|---|
| 9 | |
| 8 | |
| 7 | |
| 6 | |
| 5 | |
| 4 | |
| 3 | |
| 2 | |
| 1 | |
| $\frac{2}{3}$ (2 outs) | |
| $\frac{1}{3}$ (1 out) | |

**5.** *Limits and ERA.* Let ERA be given by,

$$E = 9 \cdot \frac{n}{i},$$

where $n$ is the number of earned runs allowed and $i$ is the number of innings pitched. Suppose that we fix the number of earned runs allowed at 4 and let $i$ vary. We get a function given by

$$E(i) = 9 \cdot \frac{4}{i}.$$

**a.** Complete the table on the left, rounding to two decimal places.
**b.** Find $\lim_{i \to 0^+} E(i)$:
**c.** On the basis of parts (a) and (b), conjecture a pitcher's earned run average if 4 runs were allowed and there were 0 outs.

REFERENCES

Adair, R. K., *The Physics of Base-ball*. New York, Harper & Row Publishers, 1990.

Baker, D., Mercer, J. and Bittinger, M. L., *You Can Teach Hitting*, Indianapolis, Bittinger Books, 1993, pp. 14 ff. www.dustybaker.com

# THE PHYSICS OF THE BASEBALL BAT

The physics of a baseball bat and of a batter's swing are the basis of a fascinating study.

## Kinetic Energy

A swing of a bat appears as a circular motion, as shown in Figure 12. A body of mass $m$ moving with a speed, or velocity, $v$ possesses a kinetic energy, $KE$, due to its translational motion which is given by

$$KE = \frac{1}{2}mv^2.$$

That is,

$$\text{Kinetic Energy} = \frac{1}{2} \cdot \text{Mass} \cdot \text{Velocity}^2.$$

The equation tells us that the velocity, or speed, of the bat is much more important than its mass, or simply stated, weight. I have done enough coaching at lower levels to know that it is virtually impossible on a "dusty" playing field to explain that increasing $v$, because it is squared, has a greater effect than increasing $m$, which is first power only.

### The Grip

In our book, *You can Teach Hitting*, Dusty and Jeff emphasize teaching the proper grip on the bat to increase speed, $v$.

There are three types of grip: the *standard*, the *modified*, and the *choke*.

With the *standard grip*, which we prefer for the young hitter, the player aligns the middle knuckles on both hands (see Figure 13A). A simple tip for setting up this grip is to lay the bat down in the fingers, not in the palms, across the calluses in the hands (see Figure 13B). The standard grip is somewhat similar to a golf grip, but without the thumbs on the bat. If the hitter lays the bat down in his fingers as though he were gripping a golf club and then picks it up, he will normally find that the middle knuckles on both hands are aligned.

The goal of the standard grip is to achieve as much quickness and speed as possible with the hands and bat. Think of "quickness" as how the bat starts. Think of "speed" as how fast it moves through the strike zone.

*Dusty Baker, Tommy LaSorda, and the Plane. Dusty Baker told me a true story about Tommy LaSorda, his manager of the Dodgers at the time. Tommy was a rather demanding and intense manager—the subject of many stories by his former players.*

*During spring training in Florida the Dodgers used a rather large commercial jet to travel to their games around the state. Tommy had the rule that if you did not get to the team plane on time after a game, you were responsible for your own transportation back to Vero Beach, their spring training base.*

*After one game, Dusty and fellow player Ken Landreaux took time to stop for cocktails. Time was fleeting, and Dusty kept prodding Landreaux to leave so they would not miss the plane and incur Tommy's wrath. But Landreaux kept procrastinating, time got away from them, and they were late.*

*Meanwhile, at the airport, Tommy was fuming and finally told the pilot to leave without Baker and friend; who arrived as the plane was taking off. They began to comment about being the object of Tommy's ire when they got back to Vero.*

*A pilot standing nearby responded to their grief, saying he had a Lear Jet and would be glad to fly them back to Vero as fast as possible. They hopped in the plane and got back to the Vero Beach airport before the Dodgers' plane. (This happened because a large commercial jet consumes more flying time because of the intricate flight patterns it must follow.)*

*Baker and Landreaux were at the gate as Tommy and the players disembarked. Off came Tommy, steaming about the missing players, only to look up and hear them say, "Hi guys! Where have you been?"*

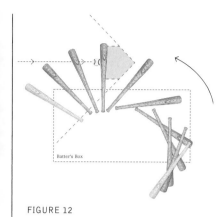

FIGURE 12
*Movements of bat through a swing.*

An important equation from the field of physics can be applied to hitting a baseball. It says that the distance a ball travels depends on the amount of energy applied to the ball when hit, and the energy applied to the ball is one half times the mass times the velocity squared; that is

$$\tfrac{1}{2} \cdot (\text{Mass}) \cdot (\text{Velocity})^2.$$

Thus bat speed has a greater bearing on how far the ball travels than does the weight of the bat. A player who is not a power hitter and not primarily concerned with distance can still make the ball come off the bat a lot harder if he develops quickness and speed. Although they may not understand why, most young hitters will notice a big improvement in bat quickness and speed if they move to the standard grip.

In the past, many major league players, such as Babe Ruth, Willie Stargell, and Dick Allen, used heavy bats. Now many power hitters in the major leagues are looking for lighter bats since most of them have been brought up using metal bats. Some even go so far as to hollow out the end of the bat. Weight of the bat should not be completely ignored—it is part of the equation for hitting the ball hard. With good technique, a hitter can be successful using a heavier bat, but learning the proper techniques is far more important than moving to a heavier bat.

Lay bat down in fingers of palms across calluses in hands.

FIGURE 13
*Gripping the bat: A. The standard grip. B. Positioning the standard grip.*

With the *modified grip*, the middle knuckles on the bottom hand align between the back knuckles and middle knuckles on the top hand (see Figure 13C). This grip tends to cause a loss in quickness and bat speed as well as a slight uppercut in the swing.

The third grip is the *choke grip*, which should not be confused with choking up on the bat. Grasping the bat in a choke grip is comparable to strangling the handle of the bat (see Figure 13D). The choke grip aligns the middle knuckles of the bottom hand with the back knuckles of the top hand. Many strong major leaguers use this type of grip, and since it does require some strength, some players aged 16 to 18 might be able to use it. The choke grip forces even more of an uppercut than does the modified grip, whether the hitter wants it or not. It also tends to cause

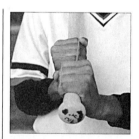

FIGURE 13
*Gripping the bat:*
*C. The modified grip.*
*D. The choke grip.*

*Finding a standard grip on a bat is similar to gripping a golf club. Be careful, however, that this does not lead you to think that swinging a golf club will enhance your baseball hitting. The swing of a baseball bat is not necessarily compatible with the swing of a golf club, even though the grips may be compatible. In fact, some major league managers have not allowed their players to play golf for this reason. Consult your hitting coach.*
— DUSTY BAKER, MANAGER, CHICAGO CUBS

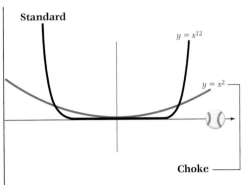

FIGURE 14
*Paths of bat using standard vs. choke grips.*

a greater loss in quickness and bat speed, particularly with younger hitters, because it tends to force greater use of the muscles in the shoulders and back rather than those in the hands, wrists, and forearms. These muscles allow for greater speed in driving the head of the bat through the strike zone.

Why do some players use the modified or choke grip instead of the standard grip? One simple answer is that often hitters are totally unaware of the standard grip. Another answer is that players simply have not been coached properly at the younger levels. They tend to grab a bat in a choking manner that seems natural to them but is in fact poor fundamental technique.

There is one more positive result of using the standard grip asserted by Dusty and Jeff. The bat swung with the standard grip stays flatter longer through the swing than the bat swung with the choke grip. As a mathematician, I explain this assertion with two graphs. The standard grip is represented by the graph of the equation $y = x^{12}$, the choke grip by the graph of the equation $y = x^2$, and the flight of pitch as a ball traveling along the *x*-axis. See Figure 14.

## Metal vs. Wooden Bats

All other variables remaining constant, a player can swing an aluminum bat faster than a wooden bat, and thus attain the advantage of extra speed.

Compare a wooden bat and an aluminum bat in terms of balance point (center of gravity, CG), and you will find the balance point of an aluminum bat farther out toward the end of the bat than the CG of a wooden bat. See Figure 15. There is more weight in the handle of a metal bat. This also gives the hitter

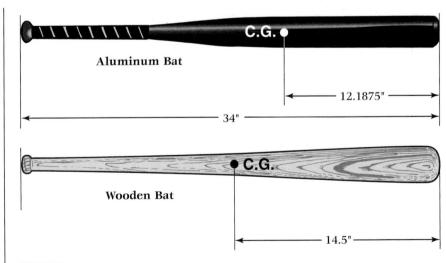

FIGURE 15
*Comparision of balance points of an aluminum bat (**A**) and a wooden bat (**B**).*

REFERENCES

I wish to thank Robert K. Adair, again, and Mark Reubers of Grove City College for their kind and helpful conversations regarding this topic.

Baker, D., Mercer, J., and Bittinger, M. L. *You Can Teach Hitting.* Indianapolis, Bittinger Books, 1993, pp. 218–220 www.dustybaker.com.

an advantage because he will hit inside pitches with more power, and will usually not break the bat above the handle, which often happens with a wooden bat. (Metal bats do break. I actually saw it happen once in a softball game.)

## Bat Vibrations and Sweet Spots

Have you ever heard a player saying "I hit the sweet spot" or "I hit that ball so well I barely felt it"? Considering the physics of a baseball bat as it hits a ball helps us understand what this means. First of all, when it is swung at a ball, a bat is a firm *rotating* object. It makes sense then that the speed and quickness of that rotation have an effect on the ball.

But after it hits the ball, the bat becomes a *vibrating* object. Although this vibration is not visible to the naked eye, the batter definitely feels it to one degree or another. To understand how the vibration takes place, take a piece of metal sheeting about 2 inches wide and 5 feet long. Hold it in the middle and shake it. The result resembles a vibrating bat (see Figure 16). Note that vibrations can be seen in the middle, where it is being shaken, and on the ends, but there are two locations where there seem to be no vibrations. These locations are called *nodes*, and locations of greatest vibration are called *antinodes*.

Although they are not visible, vibrations occur when the bat hits that ball. There are two nodes on a bat, one in the handle and one up in the barrel. Hopefully, the ball does not hit the handle node because that is where the bat is held. But the ball can hit the barrel node. That node is called the *sweet spot* (see Figure 17).

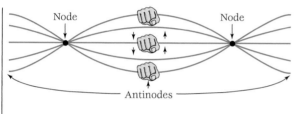

FIGURE 16
*The nodes and antinodes of a vibrating strip of metal.*

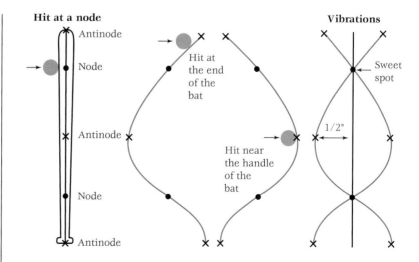

FIGURE 17
*The nodes, antinodes, and sweet spot of a vibrating bat.* (Reprinted with permission from *The Physics of Baseball* by Robert K. Adair, New York: HarperCollins Publishers, 1990, p. 91).

If the ball hits near a node, there is very little, if any vibration. If the ball hits an antinode, there is maximum vibration, possibly as much as one half inch. This explains why the greatest stinging effect occurs if a ball is hit on the antinode just above the handle. If it hits anywhere other than on a node, a vibration is created. Any vibration, no matter how slight, takes energy away from the ball, shortening the distance that it travels off the bat.

You can actually "hear" where the nodes are on a wooden bat by screwing a hook into the top end of the bat and hanging it vertically from the ceiling. Put some adhesive tape on the end of a hammer and move from one end of the bat to the other, tapping the bat with the hammer. The changes in sound will indicate the position of the nodes and places in between. The hitter could place a piece of tape on the sweet spot. The node or sweet spot on a metal bat is longer than on a wooden bat, but it cannot be discovered in the manner we just described because of the air shaft inside the bat. The longer sweet spot on a metal bat is another hitting advantage of metal over wooden bats.

## THE FORMULAS OF BILL JAMES

The Society for American Baseball Research, SABR, is an organization of baseball enthusiasts who analyze the game to its utter depths in a literary-historical-statistical manner. One of SABR's most famous members, baseball statistician Bill James, is the author of numerous analytical books dedicated to baseball. James's credibility is attested by the fact that he is now employed by the Boston Red Sox as an analyst of baseball talent. Billy Beane, general manager of the Oakland Athletics, is a proponent of James's methods, so much so that Beane now enjoys an outstanding reputation for building a quality team on a low budget. For example, one thing Beane discovered through James is that on-base percentage is three times more important than slugging percentage in determining a hitter's contributions. Another is that a team should only draft pitchers with college as well as high school experience.

### The Pythagorean Theory

I enjoy this one because it has a unique comparison to the Pythagorean Theorem for right triangles, but the analogy stops there.

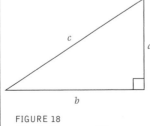

$$c^2 = a^2 + b^2 \Rightarrow \frac{c^2}{a^2 + b^2} = 1$$

FIGURE 18
*The Pythagorean Theorem.*

REFERENCES

James, B., *The Bill James Baseball Abstract.* New York, Ballantine Books, 1986. James is the author of many baseball statistical books, more recently books for fantasy leagues.

Lewis, M., *Moneyball: The Art of Winning an Unfair Game.* New York, Norton, 2003.

*Society for American Baseball Research,* P. O. Box 93183, Dept. MLS, Cleveland, OH, 44101, 1-216-575-0500.

*Bill James*

An old baseball adage says that "pitching and defense will beat offense every time." James uses what he calls a Pythagorean Theory to refute this adage. He conjectures that the ratio of a team's wins to the sum of its wins and losses (winning percentage) is approximately the same as the ratio of the square of its runs scored to the sum of the square of its runs scored plus the square of the opposition's runs scored:

$$\frac{\text{Runs}^2}{\text{Runs}^2 + (\text{Opposition Runs})^2} \approx \frac{\text{Wins}}{\text{Wins} + \text{Losses}} = \text{Winning Percentage}$$

James asserts that an improved approximation, which he found by a statistical technique, is given by

$$\frac{\text{Runs}^{1.83}}{\text{Runs}^{1.83} + (\text{Opposition Runs})^{1.83}} \approx \frac{\text{Wins}}{\text{Wins} + \text{Losses}} = \text{Winning Percentage}$$

$$\text{Standard Error} = \pm 4.15 \text{ wins}$$

**Table 9**

| NL WEST | W–L | PCT | GB |
|---|---|---|---|
| Arizona | 92–70 | .568 | — |
| San Francisco | 90–72 | .556 | 1 |
| Los Angeles | 86–76 | .531 | 6 |
| San Diego | 79–83 | .488 | 13 |
| Colorado | 73–89 | .451 | 19 |

*W = Wins, L = Losses; PCT = Winning Percentage; GB = Games behind first place team.*

For example, the final standings of the National League West in a recent year are shown in Table 9. The winning percentage of the Giants is given by

$$\frac{\text{Wins}}{\text{Wins} + \text{Losses}} = \frac{90}{162} = 0.556$$

Let's see how this compares to the Pythagorean Theory. The Giants scored 799 runs. Their opposition scored 749. Then

$$\frac{\text{Runs}^{1.83}}{\text{Runs}^{1.83} + (\text{Opposition Runs})^{1.83}} = \frac{799^{1.83}}{799^{1.83} + 749^{1.83}} = 0.530$$

The results are quite similar. Note that the latter result really has only offensive statistics on runs scored by each team; there are no pitching or defensive statistics in the formulas. James asserts that this formula is enough to show that pitching and defense are *not* more important than offense.

### The Favorite Toy Formula

Career home run records and the possibility of getting 3,000 hits are two goals of outstanding players. Is there a way to make a prediction about outstanding players reaching a very difficult goal? James has developed a way, called his Favorite Toy Formula, which is best explained by working through an example.

ALEX RODRIGUEZ.   Considered one of the most outstanding players in the game, Rodriguez is baseball's highest paid player at a salary of about $25 million per year. Can he get 3,000 hits in his career? Let's look at his hit total for all the years he has played through his MVP year, 2003. Refer to Table 10 as we work through the following five questions.

**Table 10**    Career Batting Statistics of Alex Rodriguez

| Batting | | | | | | | | | | | | | | | |
|---|---|---|---|---|---|---|---|---|---|---|---|---|---|---|---|
| YEAR | Team | AVG | G | AB | R | H | 2B | 3B | HR | RBI | BB | K | IBP | SLG | OPS |
| 1994 | Sea | .204 | 17 | 54 | 4 | 11 | 0 | 0 | 0 | 2 | 3 | 20 | .241 | .204 | .445 |
| 1995 | Sea | .232 | 48 | 142 | 15 | 33 | 6 | 2 | 5 | 19 | 6 | 42 | .264 | .408 | .672 |
| 1996 | Sea | .358 | 146 | 601 | 141 | 215 | 54 | 1 | 36 | 123 | 59 | 104 | .414 | .631 | 1.045 |
| 1997 | Sea | .300 | 141 | 587 | 100 | 176 | 40 | 3 | 23 | 84 | 41 | 99 | .350 | .496 | .846 |
| 1998 | Sea | .310 | 161 | 686 | 123 | 213 | 35 | 5 | 42 | 124 | 45 | 121 | .360 | .560 | .920 |
| 1999 | Sea | .285 | 129 | 502 | 110 | 143 | 25 | 0 | 42 | 111 | 56 | 109 | .357 | .586 | .943 |
| 2000 | Sea | .316 | 148 | 554 | 134 | 175 | 34 | 2 | 41 | 132 | 100 | 121 | .420 | .606 | 1.026 |
| 2001 | Tex | .318 | 162 | 632 | 133 | 201 | 34 | 1 | 52 | 135 | 75 | 131 | .399 | .622 | 1.021 |
| 2002 | Tex | .300 | 162 | 624 | 125 | 187 | 27 | 2 | 57 | 142 | 87 | 122 | .392 | .623 | 1.015 |
| | | AVG | G | AB | R | H | 2B | 3B | HR | RBI | BB | K | OBP | SLG | OPS |
| Totals | | .309 | 1114 | 4382 | 885 | 1354 | 255 | 16 | 298 | 872 | 472 | 869 | .380 | .579 | .958 |

*Alex Rodriguez*

Let's calculate the probability of Rodriguez getting 3,000 hits in his career.

> How old is he? Born July 27, 1965, Age 27
>
> Total number of hits = 1,354
>
> How well is he doing lately?
>
> How much longer can he play?

1.  *Distance.* How far away from the goal of 3,000 hits is he? The number of hits he needs, *N*, is given by

$$N = 3000 - 1535 = 1465.$$

2.  *Momentum.* How fast is Rodriguez approaching his goal? James answers that using a "What have you done for me lately?" weighted average given by

$$EPL = \frac{3(H_1) + 2(H_2) + H_3}{6},$$

where

> $H_1$ = Total number of hits last season,
>
> $H_2$ = Total number of hits season before last, and
>
> $H_3$ = Total number of hits three seasons ago.

In Rodriguez's case,

$$EPL = \frac{3(H_1) + 2(H_2) + H_3}{6}$$

$$= \frac{3(\text{Hits in 2003}) + 2(\text{Hits in 2002}) + (\text{Hits in 2001})}{6}$$

$$= \frac{3(181) + 2(187) + (201)}{6} \quad \text{Substituting from Table 10}$$

$$= 186.3 \text{ hits per year}$$

$$= \text{Rodriguez's Established Performance Level.}$$

3. *Time.* How much longer can Rodriguez play? That number of years, $Y$, is given by

$$Y = 24 - 0.6(\text{Present Age})$$
$$= 24 - 0.6(28)$$
$$= 7.2.$$

A player 37 years or older is given exactly 1.5 years left to play.

4. *Projected Remaining Hits.* The number of hits, $H_p$, we can predict for Rodriguez is

$$H_p = (\text{Years Rodriguez Can Play}) \cdot (\text{Rodriguez's EPL})$$
$$= (7.2)(186.3)$$
$$= 1341.36.$$

5. *Probability of Attaining 3,000 Hits.* We then compute the probability, $P$, of Rodriguez getting 3,000 hits as given by

$$P = \frac{Y \cdot EPL - 0.5N}{N}$$

$$= \frac{(7.2) \cdot (186.3) - 0.5(1465)}{1465}$$

$$\approx 0.42 \text{ or } 42\%.$$

Based on James's "What have you done for me lately?" Favorite Toy Formula, we must conclude that it is quite probable that Rodriguez will reach 3,000 hits. His youth and outstanding performance are to his advantage.

It is of interest to look at other outstanding players to see how valid this formula was in making predictions before their careers ended.

| | |
|---|---|
| Robin Yount | By 1990, age 35, he had 2,747 hits. |
| | $P = 151\%$ "$=$" 97% (97% is as high as James allows). |
| | He passed 3,000 hits. |
| Dave Winfield | By 1992, age 41 he had 2,866 hits. |
| | $P = 123\% = 97.1\%$. |
| | He passed 3,000 hits. |
| Eddie Murray | By 1992, age 37, he had 2,646 hits. |
| | $P = 53\%$. |
| | He passed 3,000 hits in 1995. |

| Andre Dawson | By 1992, age 39 he had 2,502 hits. |
| | $P = -0.3\%$ "=" 0% (0% is as low as James allows). |
| | He retired, not passing 3,000 hits. |
| Tony Gwynn | By 1992, age 32, he had 2,021 hits: $P = 33\%$. |
| | By 1996, age 36, he had 2,560 hits: $P = 44\%$. |
| | By 1997, age 37, he had 2,780 hits: $P = 84\%$. |
| | He retired after 2001, with 3,141 hits. |

BARRY BONDS.   After Barry Bonds broke the single-season home run record and four other major-league records in 2001, speculation began as to whether he could break Hank Aaron's career record of 755 home runs. Bonds was 39 years old after the 2003 season, and had a total of 658 home runs. Let's determine the probability of Bonds breaking Aaron's record.

1.  *Distance.* $N = 755 - 658 = 97$.

2.  *Momentum.*

$$EPL = \frac{3(H_1) + 2(H_2) + H_3}{6}$$

$$= \frac{3(\text{HRs in 2003}) + 2(\text{HRs in 2002}) + (\text{HRs in 2001})}{6}$$

$$= \frac{3(45) + 2(46) + (73)}{6} \quad \text{Substituting from Table 10}$$

$$= 50 \text{ homers per year}$$

$$= \text{Bond's Established Performance Level.}$$

3.  *Time.* Since Bonds was 39, his estimated years to play is fixed at 1.5 years.

4.  *Projected Remaining HRs.* The number of homers, $H_p$, we can predict for Bonds is

$$H_p = (\text{Years Bonds Can Play}) \cdot (\text{Bonds's EPL})$$

$$= (1.5)(50)$$

$$= 75.$$

5.  *Probability of Attaining 755 HRs.* We then compute the probability, $P$, of Bonds getting 755 homers as given by

$$P = \frac{Y \cdot EPL - 0.5N}{N}$$

$$= \frac{(1.5)(50) - 0.5(97)}{97}$$

$$\approx 27\%.$$

(If you ever apply this formula and get a negative answer, convert it to 0% since probability cannot be negative.) Thus, by James's Favorite Toy

**Table 11**  Probability of Bonds Breaking Aaron's Home Run Record During 2004

| Bonds Home Run Total | Probability of Breaking Aaron's All-Time Record of 755 |
|---|---|
| 658 | 27% |
| 660 | 29% |
| 670 | 38% |
| 690 | 65% |
| 700 | 86% |
| 710 | 97% |

Formula, there is a reasonable chance that Bonds will break Aaron's record. But Bonds is in a class by himself, and could even justify a change in James's formula. Bonds draws record numbers of intentional walks because opposing managers are in such awe of his hitting skills. Bonds also keeps in excellent condition, employing three trainers—one for stretching and agility, one for strength, and one as a full-time chef to tend to his muscular 6'2" body.

It is interesting to note, as Table 11 shows, how the probability of Bonds breaking the record will increase in 2004 with each subsequent home run. Once he gets past 700 home runs, the probability is 97% of his breaking the record. The only data point we can use to verify this conjecture is Aaron, who did break Ruth's record of 714. (Actually the formula gives 117%, but probability cannot exceed 97%). Keep your eyes on this record.

## Conclusion

I had the good fortune to be able to contact Bill James by e-mail and discuss with him how he developed these formulas. In the case of the Pythagorean Theory, he said he checked records of several hundred teams, including minor league teams, to figure out what the error was. He had been working on the problem so long that when he finally hit upon the solution, he "basically knew it would work even before" he tested it.

James went on to say that the Favorite Toy Formula is hard to verify because the field of players who meet the standards is so small. James reinforced that the formula can only be applied to outstanding players. James said, "It is impossible to know whether Rodriguez's chances of hitting 3,000 hits is 37% or 39% or 40%. We know it is somewhere in there. We know that if you take a group of players whose chances of reaching 3,000 hits total up to 7.04, it is very likely that 6 to 8 of them will, in fact, get 3,000 hits. We know that if you take a group of players who have about a 60% chance of hitting 500 homers, a little more than half of them will. I think this is as much as can be known about verifying the Favorite Toy Formula."

### Pursuing Further

1. Use the data for Rodriguez in Table 10 and the Favorite Toy Formula to determine the probability of Rodriguez breaking Aaron's career home run record of 755.
2. Use the data for Bonds in Table 11 and the Favorite Toy Formula to determine the probability of Bonds getting 3,000 hits.
3. Examine the data on Ted Williams's career prior to his going into World War II. Determine the probability of Williams breaking Ruth's career home run record of 714. Ted was born in 1918.

## THE MAGIC NUMBER

In September, TV and radio announcers start talking about the "magic number." Suppose the Giants are in first place and the Dodgers are in second. They might say something like the Giants' magic number is 12. That means that any combination of Giants wins and Dodger losses that total 12 will guarantee the pennant for the Giants. Each day the magic number is less than or equal to the day before. For example, if the Giants win and the Dodgers lose, the magic number goes down by 2. If both teams lose, this means that the Dodgers lost, then the magic number goes down by just 1. If the Giants lose and the Dodgers win, there is no change in the magic number. If the Giants win, and the Dodgers win also, the magic number goes down just by 1 because we consider only the win by the Giants. If there is a day where the Giants win and the Dodgers do not play, the magic number still goes down by 1. The magic number is always a whole number. When the magic number is 1, a tie is clinched; when it is 0, the pennant is clinched.

I always wondered how the magic number was computed, but never saw a formula until one appeared in the sports pages of *USA Today*, as shown in the following box.

---

**Magic Number**

The magic number, $M$, is given by

$$M = G - P - L + 1,$$

where

$G = $ the total number of games in the season, 162 for the major leagues,

$P = $ the total number of games that the first place team has played, and

$L = $ is the number of games the leading team is ahead in the loss column.

---

Suppose these are the standings:

| Team | Wins | Losses | Pct. | GB |
|---|---|---|---|---|
| San Francisco Giants | 87 | 60 | 0.591 | — |
| Los Angeles Dodgers | 84 | 64 | 0.568 | $3\frac{1}{2}$ |

Then

$$G = 162, P = 87 + 60 = 147, \text{ and } L = 4,$$

so

$$M = G - P - L + 1$$
$$= 162 - 147 - 4 + 1$$
$$= 12.$$

I was not satisfied with this formula because it did not reflect what the announcers kept saying, "any combination of Giants wins and Dodger losses that total 12 will guarantee the pennant for the Giants." I wanted to find a better formula. The development is as follows. Let

$W_1, L_1$ = Wins, Losses for the first place team, and

$W_2, L_2$ = Wins, Losses for the second place team.

Then

$$M = G - P - L + 1$$
$$= G - (W_1 + L_1) - (L_2 - L_1) + 1$$
$$= G - W_1 - L_1 - L_2 + L_1 + 1$$
$$= G - W_1 - L_2 + 1$$
$$= G - (W_1 + L_2) + 1.$$

This formula reflects what the announcers were saying. You can apply it easily when reading the sports pages; add the wins of the first place team to the losses of the second place team, subtract from 162, and add 1.

    This was an MPX for me because I was able to make this deduction on my own, even though the announcer's statisticians may have been using the same method.

*I have had the joy of meeting Willie Mays and Duke Snider, but not Mickey Mantle, now deceased.*

MPX MUSIC: "The Final Game," from the movie *A League of Their Own*, Hans Zimmer, arr. Steven Reineke, Avon Gate Music (BMI). Track 12 on the CD *Play Ball!* by the Cincinnati Pops directed by Erich Kunzel.
MOVIE: *A League of Their Own* (1992) starring Tom Hanks and Geena Davis.

    Another way to understand the magic number is to look again at the Giants situation. What is the most number of games the Dodgers can win?

$$162 - 64 = 98 = G - L_2.$$

Then how many games must the Giants win to exceed the number of wins that the Dodgers can get?

$$98 - 87 + 1 = 12 = (G - L_2) - W_1 + 1.$$

So, 12 is the magic number.

    We can create another proof with variables by reasoning from this concrete example to the abstract.

    What is the most number of games the second place team can win?

$$G - L_2$$

Then how many games must the first-place team win to exceed the number of wins that the Dodgers can get?

$$M = (G - L_2) - W_1 + 1$$
$$M = G - (W_1 + L_2) + 1$$

The answer, $M$, is the magic number.

### Pursuing Further

Compute the magic number for each first-place team, given the standings.

| | | W-L | PCT | GB |
|---|---|---|---|---|
| **1.** | AL WEST | | | |
| | Seattle | 88–64 | .579 | — |
| | Oakland | 84–67 | .556 | $3\frac{1}{2}$ |
| **2.** | AL EAST | W-L | PCT | GB |
| | NY Yankees | 102–58 | .638 | — |
| | Boston | 101–59 | .631 | 2 |
| **3.** | NL EAST | W-L | PCT | GB |
| | Atlanta | 96–43 | .691 | — |
| | Philadelphia | 72–66 | .561 | $13\frac{1}{2}$ |
| **4.** | NL CENTRAL | W-L | PCT | GB |
| | Houston | 90–37 | .709 | — |
| | St. Louis | 89–42 | .679 | $5\frac{1}{2}$ |

MARV BITTINGER
DUSTY BAKER CD-ROM    1996

1996    Marv Bittinger

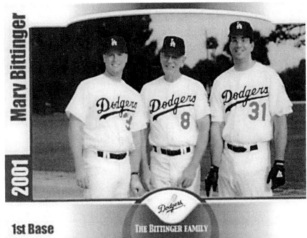

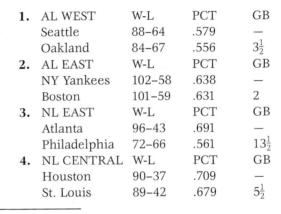

*These are three of Marv's pseudo baseball cards when he attended the San Franciso Giants baseball camp in 1996, the Los Angeles Dodger Adult baseball camp later that year, and the third with his two sons Chris (left) and Lowell (right) when they attended the Los Angeles Dodger Adult baseball camp in November of 2001.*

# The Humorous Side of My Journey

# 4

**INTRODUCTION**

As a student, I thoroughly enjoyed having an instructor bring some humor to a math lecture. Occasionally, some subtle humor can be found in my own math books, but such instances are few and far between. Over the years, I have gathered some humorous quips and stories from books, journals, math classes, and other instructors, even inventing a few on my own on occasion. My favorites have become part of a talk, called "Humor in Mathematics," that I have often presented at educational conferences. The attendees seemed to enjoy the levity the talk brought to the meetings.

I've always wanted to include more humor in my math books, but inevitably such inclusion fails to appeal to my audience. I believe humor lets students know you are human, but too much humor can compromise the book's integrity. If we had a joke or a cartoon on every page, the instructor would deem the book unprofessional and drop it, even though the subject matter might be superb.

One attempt at humor involved the use of funny names in word problems. The idea lasted for about two editions. If a problem was on working to build a brick wall, we might say that "Red Bryck" was working on the wall. If it was on carbon dating, we might name a chemist, "Ray Dioactive." In another we said "R. U. Teed" paid $30.75 for a dozen golf balls, and in another "One year Tim Burr, a logger, earned $87,450."

Mike Keedy and I had lots of fun thinking up those silly names and to this day they do not bother me, but the marketplace had a different reaction, especially after one edition. Then the names grew old and were no longer funny to the instructor, even though they were still fresh for the next students. But the reader must keep in mind that mathematicians have a sense of humor, too—even if students don't appreciate it.

This is not a textbook, so I will take the liberty of sharing some mathematical humor I have enjoyed. It is a collage of humor of all types from real life or stories I've heard or, like the following, from the imagination.

*Jonathan Winters in his Maudie Frickett costume.*

## MAUDIE FRICKETT, THE WORLD'S OLDEST MATH TEACHER

I had yet to write any school division–level books when I often helped Walt Marsh, a K–12 textbook salesman for Addison-Wesley, do workshops for books after they had been adopted in various school systems around Indiana. At a sales meeting, Walt did the following impersonation of a character, Maudie Frickett, in the repertoire of comedian Jonathan Winters.

If you do not recall Maudie, let me tell you that she was a feisty old lady, who could bite you with her quick wit. Imagine here that Maudie has been an elementary teacher of mathematics for some 79 years, and she is frustrated with innovations in mathematics education. Maudie (M) is being interviewed by a young, eager announcer (A) doing a news feature story about mathematics education. I laughed till I cried when I heard Walt deliver this impersonation. It went something like this, with some editing on my part:

*A:* What is your name, madam?

*M:* My name is Maudie Frickett.

*A:* Is that Miss or Mrs. Frickett?

*M:* Msssssss Frickett! I'm into this women's lib, Sonny.

*A:* How long have you been teaching, Ms. Frickett?

*M:* I have been teaching first grade for 79 years, right here in this room, except for 3 years in the Marine Corps in World War II.

*A:* Did your military experience help you to become a better teacher in any way?

*M:* Yes, it sure did! I learned how to make the little "brats" line up a whole lot straighter! It comes in handy at recess time.

*A:* Now, Ms. Frickett, I want to talk to you about mathematics. You do teach it, don't you?

*M:* Well, I don't know about mathematics, but I do teach the children their numbers.

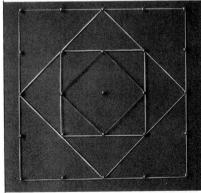
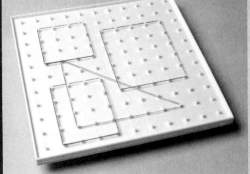
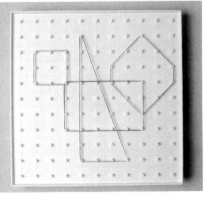

*Geoboards.*

*Trivia Question: Who is probably the only person in the world to shake hands with the first man to fly, Orville Wright, and the first man to walk on the moon, Neil Armstrong? Answer: Jonathan Winters. JW was born and raised in Dayton, Ohio, home of Orville Wright. Neil Armstrong was from nearby Wapakoneta, Ohio.*

*A:* Ms. Frickett, how do you handle a missing addend in the first grade?

*M:* Handle it? I skip the son of a gun—that's how I handle it.

*A:* Then you don't believe in teaching related sentences?

*M:* Well, I think prisoners should serve all the time of the sentences they receive and not have them related.

*A:* Well, what do you think is important, Ms. Frickett?

*M:* Facts, facts, facts, that's what! The children can't get nowhere without their facts. I pride myself in children knowing facts to 20. Drill, drill, drill—that's my motto! This is another trick I picked up in the Marine Corps.

*A:* How do you motivate the children to learn their facts?

*M:* I stand behind them with a big ruler. When they make a mistake, I touch them with a sharp one, right on the mastoid bone. That's my motivation.

*A:* Ms. Frickett, what is your opinion of the associative principle?

*M:* Why, I've got another name for that *idiot* than associative. Why, I said to him the other day. . . .

*A:* No, no, no, Ms. Frickett, not the principal of the school, the associative principle in mathematics.

*M:* Oh, well I hadn't thought that much about it, to tell the truth.

*A:* Do you ever use a geoboard, Ms. Frickett?

*M:* Yes, indeed, I do. Instead of a ruler, I sometimes use a geoboard. If they make one mistake, they get the flat side, and two mistakes, I whomp them with the nails.

*A:* Do you ever use concrete objects in your teaching?

*M:* Well, I prefer to whomp them with a ruler or a geoboard, but I bet a concrete block would really get their attention.

*A:* How do you feel about individualized instruction?

*M:* Well, this is how I do it. If they fail the test the first time, I lock them in a closet with no food until they do pass a retest. It is amazing how they come around after that.

*A:* What do you think about the use of calculators?

*M:* Buttons and batteries, they will fail you everytime. I prefer drill and punishment.

*A:* I presume you have been teaching the metric system?

*M:* The what?

*A:* The system of measurement that is used throughout the rest of the world.

*M:* Oh, that evil plot. I say our God-given quarts, inches, and pounds is good enough! Let the rest of them use our system!

*A:* How many children do you have in your class this year, Ms. Frickett?

*M:* Well, I started with 26, but for some reason I now have only 3 children. But these 3 sure sit at attention when I speak. Parents just don't know how to raise well-adjusted children these days.

*A:* Ms. Frickett, you are a disgrace to education. You should be fired!

*M:* Hee, hee, no way! I've got tenure, Sonny, which is more than you have.

*A:* What more can I say? Thank you, Maudie Frickett!

*M:* Thank you, Sonny. Come again.

Walt did this impersonation in a black dress and gray wig. He later became National Sales Manager for Addison-Wesley's School Division.

## HUMOR AMONG GENIUS

The following stories of four famous mathematicians are related with the permission of their publishers. I include them because each story struck a humorous chord with me, and in some small way I found myself in them.

### Edward Kasner

Kasner (1878–1955) was co-author with James R. Newman of the very popular book, *Mathematics and the Imagination*, originally published in 1940 and revised in 1989 by the Cobb Group. Kasner once asked his young nephew to invent a name for a very large number. The boy called it a *googol*, and Kasner defined googol in decimal notation as a 1 followed by 100 zeros; that is, a googol is $10^{100}$. It is interesting to note that $10^{80}$ is considered to be the total number of elementary particles in the entire universe, so a googol is quite adequate for describing our "existence," so to speak. Kasner went on to define a *googolplex*, as $10^{\text{Googol}}$, or $10^{10^{100}}$. The question, "What is a 1 followed by 100 zeros?" was asked on the British version of the "Who Wants to Be a Millionaire?" TV quiz show. It was answered correctly.

Clifton Fadiman tells two stories about Kasner, which I really enjoyed and include here.

*He (Kasner) was a small man, with a gentle voice, but deceptively so, for hidden within it lay an overtone of tolerant irony. I never knew him to wear anything but a pepper-and-salt suit, winter and summer, and I do not believe he owned more than one. He had long ago discarded the belt as an unnecessary modern invention and spent a good deal of his time holding up his trousers. At his lectures his students were often torn between conic sections and equally abstruse mathematical calculations as to whether or not Dr. Kasner's pants would fall down before the lecture period's elapsed time. His lectures were thoroughly unconventional and he was one of the greatest teachers I have ever known. (p. xxiv–xxv)*

• • •

*Once, at an informal evening attended by half a dozen logicians and mathematicians, one of the latter ventured the opinion that to become a good mathematician you should not have good teachers, as they prevented you from learning by yourself. He then added, courteously, "I was lucky. I had only one great teacher—Dr. Kasner here." Kasner looked up; his eyes twinkled; he said mildly, "I had none." (p. xxvii)*

*Edward Kasner, mathematician, (1878-1955).*

REFERENCES

Fadiman, C. *The Mathematical Magpie*, New York, Simon & Schuster, 1981, pp. xxiv–xxvii.

I could not disagree more with the latter statement, especially since I was blessed to have so many excellent teachers, but there might be an element of truth there, given that you have an infinite amount of time to learn.

## Norbert Wiener

Norbert Wiener (1894–1964), a precocious mathematician, received his Ph.D. from Harvard University at the age of 18. He later studied under such great mathematicians as Bertrand Russell, G. H. Hardy, and David Hilbert. Wiener contributed some mathematical papers of great importance in areas such as Brownian motion, applications of Fourier transforms, and cybernetics. Hans Freudenthal, a great mathematician also, once said of Wiener, "He was a famously bad lecturer."

One story about Wiener goes as follows:

*After a long and very abstruse lecture on a particular problem he reached the end and underlined the solution. A bright student popped up and asked Wiener, "Isn't there another solution?" Wiener paused and thought for awhile. Then he said, "Yes, there is." He walked up to the blackboard and underlined the solution a second time.*

George Polya tells the following amusing story about Wiener:

*Now here is the story which was widely told, but is hardly true. It is about a student who had a great admiration for Wiener, but never had the opportunity to talk to him. The student walked into a post office one morning. There was Wiener, and in front of Wiener a sheet of paper on the desk at which he looked with tremendous concentration. Suddenly Wiener ran away from and then back to the paper, facing it again with tremendous concentration. The student was deeply impressed by the prodigious mental effort mirrored in Wiener's face. He had just one doubt; should he speak to Wiener or not? Then suddenly there was no doubt, because Wiener, running away from the paper, ran directly into the student who then had to say, "Good morning, Professor Wiener." Wiener stopped, stared, slapped his forehead and said: "Wiener, that's the word." (p. 748)*

Somehow I have the nagging fear that someday I will be capable of a similar incident.

## David Hilbert

Hilbert (1862–1943) was a brilliant mathematician, famous for his work on infinite dimensional spaces, later called Hilbert spaces, a concept useful in mathematical analysis and quantum mechanics. He also contributed to many other branches of mathematics, including algebraic number fields, invariants, functional analysis, in-

---

*Norbert Wiener, mathematician (1894–1964).*

REFERENCES
Polya, G. "Some Mathematicians I Have Known," *The American Mathematical Monthly*, Vol. 76, No. 7, pp. 746–753, 1969.

*A professor is one who can speak on any subject, for precisely fifty minutes.*
—NORBERT WIENER

*The advantage is that mathematics is a field in which one's blunders tend to show very clearly and can be corrected or erased with a stroke of the pencil. It is a field which has often been compared with chess, but differs from the latter in that it is only one's best moments that count and not one's worst. A single inattention may lose a chess game, whereas a single successful approach to a problem, among many which have been relegated to the wastebasket, will make a mathematician's reputation.*
—NORBERT WIENER, in *Ex-Prodigy: My Childhood and Youth*

REFERENCES
Polya, G. "Some Mathematicians I Have Known," *The American Mathematical Monthly*, Vol. 76, No. 7, p. 747, 1969.

*David Hilbert, mathematician (1862–1943).*

We must know. We will know.
—DAVID HILBERT, in a speech in Königsberg in 1930. This was written on his tombstone

No other question has ever moved so profoundly the spirit of man; no other idea has so fruitfully stimulated his intellect; yet no other concept stands in greater need of clarification than that of the infinite.
—DAVID HILBERT, quoted in E. Maor, To Infinity and Beyond. Boston, Birkheauser, 1987.

A mathematical theory is not to be considered complete until you have made it so clear that you can explain it to the first man you meet on the street.
—DAVID HILBERT

This quote, in my opinion, is the most challenging in mathematics education.

If I were to awaken after having slept for a 1,000 years, my first question would be: Has the Riemann hypothesis been proven?"
—DAVID HILBERT

tegral equations, mathematical physics, and the calculus of variations. In 1899 he published *Grundlagen der Geometrie*, a major influence in promoting the axiomatic approach to mathematics, which has been one of the major characteristics of the subject throughout the twentieth century.

Hilbert liked to pose and work on challenging problems such as the continuum hypothesis, the well ordering of the real numbers, the transcendence of powers of algebraic numbers, and the extension of Dirichlet's principle. Two of the problems he worked on, Goldbach's conjecture and the Riemann hypothesis, remain unsolved as of the writing of this book. The Riemann hypothesis has to do with the distribution of the prime numbers, but is very difficult to state. On the other hand, Goldbach's conjecture is very easy to state: *Every even integer greater than 2 can be written as the sum of two primes*.

For example, $4 = 2 + 2$, $10 = 3 + 7$, and $98 = 19 + 79$, all illustrate this conjecture. Despite lots of progress, this remains unproved.

George Polya tells the following amusing story about Hilbert:

*First about absent-mindedness. Many such stories are told about Hilbert. Are they true? I doubt it, but some are quite good. There is a party in Hilbert's house and Frau (I mean Mrs.) Hilbert suddenly notices that her husband forgot to put on a fresh shirt. "David," she says sternly, "go upstairs and put on another shirt." David, as it befits a long married man, meekly obeys and goes upstairs, yet he does not come back. Five minutes pass, ten minutes pass, yet David fails to appear and so Frau Hilbert goes up to the bedroom and there is Hilbert in his bed. You see, it was the natural sequence of things: He took off his coat, then his tie, then his shirt, and so on, and went to sleep. (p. 747)*

This amused me so much when I first read it that I told the story with glee to my family and friends, and continue to do so in my frequent speeches on humor in mathematics. My amusement was somewhat tempered when Elaine once suggested that I go upstairs and change my clothes to go out to eat. I did come back downstairs, but I had my pajamas on.

*G. H. Hardy, mathematician.
(1877–1947).*

## Godfrey Harold Hardy

Godfrey (1877–1947), a British mathematician, published *A Course of Pure Mathematics,* in 1908. It was the first rigorous exposition of number, function, and limit adapted to the undergraduate. Hardy's first student, Burkill, said of this book, "It transformed university teaching." Hardy developed a law describing proportions of dominant and recessive genetic traits, which proved to be of major importance in blood group distribution. Probably Hardy's most famous work was *A Mathematician's Apology*. Written in 1940, it is considered one of the most vivid descriptions of how a mathematician thinks and of the pleasure of mathematics.

Like many mathematicians, Hardy was eccentric. He disliked having his picture taken and his first action when entering a hotel room was to cover the mirror with a towel.

Polya tells this paradoxical story about Hardy, which illustrates his eccentricity:

*Hardy, an atheist, stayed in Denmark with Bohr (another mathematician) until the very end of the summer vacation, and when he was obliged to return to England to start his lectures there was only a very small boat available (there was no airplane traffic at the time). The North Sea can be pretty rough and the probability that such a small boat would sink was not exactly zero. Still, Hardy took the boat, but sent a postcard to Bohr: "I proved the Riemann hypothesis. G. H. Hardy." You're not laughing? It is because you don't see the underlying theory. If the boat sinks and Hardy drowns, everybody must believe that he has proved the Riemann hypothesis. Yet God would not let Hardy have such a great honor and so he will not let the boat sink. (pp. 752–753)*

*Young men should prove theorems, old men should write books.*

—G. H. HARDY, quoted by Freeman Dyson in "Mathematician, Physicist, and Writer," interview with D. J. Albers, *The College Mathematics Journal,* Vol. 25, No. 1, 1994.

REFERENCE
Polya, G. "Some Mathematicians I Have Known," *The American Mathematical Monthly,* Vol. 76, No. 7, p. 752–753, 1969.

*. . . whereas most people are so frightened of the name of mathematics that they are ready, quite unaffectedly, to exaggerate their own mathematical stupidity.*

—G. H. HARDY, A Mathematician's Apology, London, Cambridge University Press, Reprinted 1998.

## TIDBITTS OF HUMOR

Over the years I have gathered the following so-called Tid"Bitts" of humor, some from the work of others, some from my very own imagination.

REFERENCE
Chin, P., Baker, K., et al., "Gentle Genius," *People Magazine,* Vol. 53 Iss. 8, 2 February 2000 p. 52.

## PEANUTS®

Charles M. Schulz' widely read cartoon, PEANUTS®, expresses a whimsical philosophy through the eyes of children and animals. His main character,

Charlie Brown, is usually victimized, a downtrodden, puzzled boy for whom things just do not seem to work out the way he expects. Other characters are Lucy, his bossy, know-it-all, cruel, friend; Snoopy, a romantic, self-deluded beagle, who thinks he is either a renowned author or a World War I fighter pilot; and Linus, who has an intense interest in philosophy and religion, as did Schulz, who was a strict Christian. He was never a particularly good student. Not only did he flunk math, but also physics, English, and Latin. I have always enjoyed his work and in his memory I'm pleased to present a few of his math-related cartoon strips.

*Charles M. Schulz, 1922–2000.*

1. Help Lucy with her dilemma. Translate Lucy's problem to a system of equations. Then solve the problem.

2. Check the factoring in Lucy's polynominal.

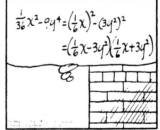

**3.** Has the denominator been rationalized correctly?

**4.** A googol is a 1 followed by one hundred zeros. Has a googol been represented in the cartoon?

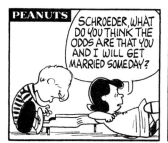

**5.** Has a student ever said this to a math teacher?

Other Peanuts® cartoons can be found on the Internet. Check out the site www.csun.edu/~hcmth014/comics/l-p/peanuts15.html.

## Quickies

1. A classroom sign once read, "Mathematics dispensed here, bring your own container."
2. Why did the student eat his math test? Because the instructor said it was "a piece of cake."
3. What do you get if you divide the circumference of a jack-o-lantern by its diameter? Pumpkin $\pi$.
4. When a student asked her grandmother for help finding the least common denominator, the grandmother replied, "You mean they haven't found that yet? They were looking for it when I was in school!"
5. Calculus students are sometimes clueless. They think that General Calculus was a war hero. If he did actually exist, he probably knew how to "integrate" his troops and "differentiate" between his allies and his enemies.

## Brain Stretchers

You will need to stretch your mind to get the "correct" answer.

1. Write as a single logarithm: $\ln A + M$ int
2. How did $a^3$ answer the question from $a^4$, "Will you do me a favor?"
3. What does $(ap_\pi)^t$ represent?
4. Does it follow that if $\lim\limits_{x \to 0} \dfrac{8}{x} = \infty$, then $\lim\limits_{x \to 0} \dfrac{5}{x} = \backsim ?$
5. Algebraic manipulations like $5x \cdot 7y = 35xy$ provide a basis for manipulations with dimension symbols. Thus,

$$(5 \text{ lb})(7 \text{ ft}) = 35 \text{ ft-lb} = 35 \text{ foot pounds}.$$

   Find each of the following:
   a. (3 barns)(4 dances)
   b. (5 dances)(3 dances)
   c. (Ice)(Ice)(Ice)
   d. (72 rocks)(8.5 bands)
6. What about $2L - 2L$ pertains to Christmas?
7. What is the Biblical significance of the geometric figure?

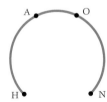

What is the meaning of each of the following?

8. | Sand |

9. $\dfrac{\text{Man}}{\text{Board}}$

10. $\dfrac{\text{Stand}}{\text{I}}$

11. |READING|

12.
R
R O A D
A
D

13. $\dfrac{0}{\text{B.A.}}$
M.S.
Ph.D.

14. $\dfrac{\text{Knee}}{\text{Light}}$

15. (Dice, Dice)

16. **NOTATION**

17. $\dfrac{\text{Ground}}{\text{Feet}}$
Feet
Feet
Feet
Feet
Feet

18. $\dfrac{\text{Mind}}{\text{Matter}}$

19. ECNALG

20. USERMANL

Simplify.

21. $\displaystyle\int \dfrac{d(\text{Cabin})}{\text{Cabin}}$

22. $3\displaystyle\int (\text{Ice})^2 d(\text{Ice})$

23. $2a\displaystyle\int (\text{Real})\,d(\text{Real})$

### A Funny Math Test

1. If $x^2$, $x^3$, $x^{-5}$ represent exponents, what do $y^2$, $y^{-4}$, $y^8$ represent?
2. What is

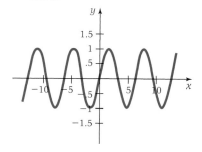

**3.** Describe

**4.** Describe

**5.** "There is three errers in this sentence." Find them.

**6.** Give the wrong answer to the question, "Which of the following two numbers is not divisible by two: 76, 84?"

In 7–11, find the common *mathematical* property.

**7.** Afghanistan navy, Juilliard School of Music Football team, arctic penguins, peacock eggs, and the real number solutions of $e^x=0$.

**8.** Barry Manilow records, Thanksgiving dinners, Rotary telephones, 8-track tapes, and numbers of the form $\sum_{k=1}^{n}(2k-1)$.

**9.** Big Foot, $\sqrt{-1}$, Harry Potter, fountain of youth, pot of gold at the end of the rainbow, UFOs.

**10.** Michigan, Hawaii, Hyperbola, Kentucky, New York.

**11.** TV time in New York from 8–11 PM, TV time in Chicago from 7–10 PM, filet mignon steaks, and natural numbers that have exactly two divisors.

Don't give the obvious answer in 12–23.

**12.** If A can saw a log in 3 hr, B in 2 hr, and C in 1 hr, why don't they just let C do it?

—FROM THE TELEVISION COMEDY SHOW, *LAUGH-IN*.

**13.** Two math teachers went out for dinner. What did they order for dessert?

**14.** A statistical study found that in a certain town, 37.2 out of every 10,000 people died of heart failure. What does this mean?

**15.** It has been discovered in a scientific experiment that in Alaska a special value of $\pi$ has to be used in calculations involving circles because of the extremely cold temperatures. What name do they give to this special value of $\pi$?

**16.** If 2's company and 3's a crowd, what are 7 and 8?

**17.** What is a square root?

**18.** What is a binomial?

**19.** What was the score of the Strontium University vs Carbon College basketball game?

**20.** Do you spell your name with a V Herr Wagner?

**21.** What was the world's first telephone number?

**22.** What do the real number system and a moron have in common?

**23.** A story going the rounds concerns three pregnant squaws who slept on animal skins—one on an elk skin, another on a buffalo skin, the third ona hippopotamus skin. The first squaw had a son; the second, a son; and the third, twin boys. What does that prove?

FROM THE READER'S DIGEST

**BONUS QUESTION**

**24.** Study the following statement and write a 100-word essay discussing it.

"It has been shown that 5 out of 4 Americans struggle with fractions!"

### The Chinese Hunger Theorem

This tongue-in-cheek story is told by William Dunham, Professor of Mathematics at Muhlenberg College in Allentown, Pennsylvania.

*But by far the most memorable teacher of my undergraduate days, and the one who had the most impact on my career, was the well-known number theorist and haberdasher, Professor Natalie Attired. I vividly recall her greatest lecture—a very clear and satisfying proof of the Chinese Remainder Theorem. I must admit, however, that an hour after she proved it I was again hungry for more mathematical knowledge.*

### Instructors, Beware!

At LAX Airport today, an individual, later discovered to be a community college math student, was arrested trying to board a flight while in possession of a Bittinger textbook, 13 kinds of textbook supplements, paper, pencil, compass, protractor, graphing calculator, and a willing mind. Authorities believe he is a member of the notorious Al-Gebra movement. He is being charged with carrying weapons of math instruction.

### Card Tables Everywhere

Dear friend Jim Biddle tells an interesting story about his major professor, Wolfgang Kappe, and his wife, Luis-Charlotte, both of whom are research mathematicians. It seems they are obsessed with not losing any creative ideas as they move about the house. So much is their obsession that all over the house they have placed card tables and chairs with paper and pencils. As they move about the house and get an idea, they sit down and write it out.

### The Frozen Yogurt Server

This is a story of mathematics gone somehow wrong. It actually happened during a convention I attended in Orlando in the fall of 2002. I endeavor to live on a heart-healthy diet and try to eat low-fat frozen yogurt instead of ice cream for dessert. But finding frozen yogurt can require extra effort when I travel.

The first day, I went to an ice cream stand in a nearby mall, which did carry frozen yogurt and bought a three-dip waffle cone for $5.08. On the second day, I went again and bought a three-dip waffle cone for $5.08. Needless to say, when I went the third day, I was programmed to expect to again pay $5.08. But, this time there was a new server. After he'd served me my three-dip cone, he'd went to the cash register and announced that the charge was $9.14. Knowing from recent experience he was wrong, I told him I had bought the same kind of cone the past two days for $5.08. He checked again and insisted the charge was $9.14. I looked around for a supervisor, and seeing none, I said I wasn't about to pay $9.14 for a three-dip cone.

Just then he looked to the end of the counter and saw a three-tiered stack of plastic cups, priced according to the number of dips in each cup, and the price for a three-dip cup was $3.79. He said, "Sir, you can buy a three-dip cup for $3.79." Now check my reasoning here:

$$\begin{array}{r} \$9.14, \text{ the price of a three-dip waffle cone} \\ -\ \underline{3.79}, \text{ the price of a three-dip cup} \\ \$5.35 \end{array} \qquad (1)$$

This meant that the value of a three dip waffle cone was $5.35 more than the value of a three-dip cup. I wondered perhaps, if the "dips" were different. Keeping in mind that I was more interested in the yogurt than the container and quite pleased with $3.79 as opposed to $5.08, I had become weary of the encounter and immediately accepted.

What did he do? He took the three dips out of the waffle cone, put them in the three-cone cup and charged me $3.79. This meant that the price of an empty waffle cone versus an empty three-dip plastic cup was also ludicrous. I knew the whole deal was not correct, but I decided it was fruitless to argue. I paid and walked away, then noticed a supervisor across the way. I thought that the server needed to learn, so I raised the issue of the price of a three-dip waffle cone with the supervisor. He checked and sure enough, the price of a three-dip cone was $5.08. I left confident that the cause of true mathematics had been upheld, plus I had a three-dip cup for $3.79. I was tempted to give the server a copy of my *Basic Mathematics* book, but decided I might get into another argument—over the price of a free book!

### The Cigarette Butts Problem

In an attempt at humor, and because it contained some good mathematics, I placed the following problem in the sixth edition of my *Basic Mathematics* book. It lasted for a few printings, until an editor found it and took it out. Stop and work it out on your own before reading beyond the problem.

*Butts are bad!*

*A person addicted to cigarettes but without cash figures out that there is enough tobacco in four cigarette butts to make one new cigarette. How many cigarettes can the person make from 29 cigarettes butts? How many butts are left over?*

If you jump to the quick conclusion, divide 29 by 4 and get 7 cigarettes with 1 butt left over, you have not thought the problem out carefully. From the 7 "new" cigarettes you get 7 more butts after they are smoked. Put the 7 butts with the 1 butt, and you get 2 more cigarettes. Smoke these, and you have produced 7 + 2, or 9 cigarettes with 2 butts left over. Now, I must admit that when you talk about this problem and repeat a certain word, over and over, you just can't help but laugh. Indeed, my co-author Judy Beecher and I laughed about this problem until our sides split. But our editor at the time was grossed out about the problem, and insisted we remove it from the book. Maybe I will try to slip it into the tenth edition since that editor has moved on. I wonder if the frozen yogurt server could do this problem?

## PROFESSOR CEDRICK R. BUNGLES

There have been mathematicians whose arrogance, poor teaching, and lousy people skills have appalled me. That these kinds of people exist in the mathematical community is one reason why the number of mathematically capable people in our society is diminishing. Rather than discuss them by name, I created a composite figure, with the fictitious name of Professor Cedric R. Bungles. Each story regarding Bungles actually happened to me in one form or another, bringing an element of humor or pathos to each situation.

### Visit to a University

Four or five times a year, I have the occasion to visit college campuses. I may be giving a talk or gathering research and information about a text I'm revising, or sometimes the faculty will invite me when they are considering one of my books for adoption. The visits never occur often enough to tire me, but I never want to visit a department unless the instructors desire and appreciate my visit.

Such was not the case when I visited the math department at Torment University, where Bungles was a professor. Typically, the Addison-Wesley sales representative might get some snacks and set them up in a department room so I can talk to people at their convenience. On this particular morning, everyone had passed through and I was able to discuss various issues regarding my book.

The last to arrive was Bungles. He walked in and slammed my book down on the table, saying loudly and arrogantly, "Dr. Bittinger, I just want you to know, I hate your book!" The AW rep turned eight shades of red at the

*Many people who want to be writers don't really want to be writers. They want to have been writers. They wish they had a book in print.*
—JAMES MICHENER

same time I did. As an author, I have learned many times that not everyone will find your work to their liking, but there is a kind and gentle way to make your point. Do you treat a guest who has traveled over 1100 miles with such a hostile comment? *I think not.*

I somehow gathered my wits, and asked what he disliked specifically. It might have been more appropriate for him to start out with some positive comments, and then say, "Now, I do have some issues to discuss with you." I would graciously have accepted them, and indeed may have picked up some helpful hints for a future revision. Instead, I left Torment University with a bad taste in my mouth.

I have often speculated on what might have been working on Bungles that day. Perhaps he had been denied tenure or harbored some envy about my success. Who knows? Perhaps Bungles had wanted to write a book himself, but had never gotten around to it. Such painful experiences go with the territory for an author.

### An Instructor Who Should Stick to Research

On another occasion, I was invited to teach a summer session at Anguish University far from home in Indianapolis. I also planned to take an analysis course during that time just to brush up on my mathematics, and Bungles, unfortunately, was my professor for that course.

Bungles had a severe stuttering problem. I harbor no prejudice toward people with such a problem, but it was virtually impossible to learn from Bungles' lectures. Since there was no textbook for the course, Bungles gave us handouts of notes each day. One day he handed out a set of notes and then gave his lecture. The next day he handed out another set of notes—which were corrections to the first day's notes. Then, lo and behold, on the third day, he handed out a correction sheet for the correction sheet. At this point my fellow students and I were lost, and unable to understand the third day's notes any more than the first day's. We expected another sheet the next day but, mercifully, none appeared. There are research mathematicians whose lectures are so superb that no textbook is needed, but not in Bungles' case.

To me Bungles also showed a very condescending attitude. At first, as I normally do, I asked lots of questions. But after getting those, "You have to be stupid to ask that!" responses, I soon gave up and slumped in my chair. It was clear that I was pond scum in his eyes. To cap it off, would you believe that Bungles happened to be teaching in the same classroom immediately after I taught my course? One day he came in and criticized me for not properly erasing the blackboard before he taught.

I was glad when the summer was over, but I grieved for his future students. In my opinion, Bungles should never have been hired as a teacher in the first place. A university professor who is employed to do any kind of teaching should have that skill carefully evaluated. Bungles might have served the university well as an administrator or as a full-time research mathematician.

## A National Reputation, No People Skills

On another occasion, I ran into Bungles at Misery University, where he was a research mathematician of national renown and the editor of one the most prestigious mathematical journals in the country.

As evidenced in other parts of this book, I have never been afraid to speak up and ask questions. So, since I had written my master's thesis on a logic topic that I thought might make a nice publication in Bungles' journal, I went to his office to inquire. He sat in his chair, staring at me, moving his eyes from head to toe and back up again, much like some men look at women. Then he said, "Who are you?" I proceeded to put my tail between my legs, figuratively, and crawl out of the room. Perhaps my thesis was worthy of publication, and his arrogance prevented my ever finding out. More likely, I was way out of my league with the question. In such a case, couldn't there have been a kinder, gentler way of saying I did not have the reputation to be considered?

*Despite what some of my friends advised, I couldn't be as unkind to him as he was to me. I have learned since that Bungles has been rude and cruel to publishing representatives and editors at other mathematics conventions.*

As sometimes happens, the tables were turned a few years later. Bungles wrote a calculus book, but it was not particularly successful and was passed from one publisher to another over the years. During this time he also wrote an algebra-trig series. Bungles and I happened to be at the same math meetings when I had just completed a revision of my algebra-trig series. I was in the Addison-Wesley booth when Bungles stopped by to check out our new series. I could sense the envy. It was all I could do not to walk up to him and say, "Who are you?"

# People Who Have Touched My Mathematical Life

# 5

INTRODUCTION

One reason I became a mathematics educator was the impact of an amazing string of incredible math teachers from seventh grade through my undergraduate years at Manchester College (MC). Another reason is the caring friends I have encountered through my journey in mathematics. I venture to say that I would not have had the success I have without the inspiration of their instruction and the influence of their lives. My success as a teacher, as an author, as a devoted husband and father, and as a friend to others are because of these people, and it is my turn to honor them in this book. Perhaps, if you have an interest in math education, you will find inspiration in these stories.

## JOHN K. BAUMGART

The math instructor that stands tall at the top of the list is Professor John K. Baumgart. John was born in Maywood, Illinois, and attended Wheaton College, where he received a bachelor of science degree in 1936. He took his master's degree at the University of Michigan in 1939. Before arriving at MC, he taught at Cumberland College in Kentucky and at Elmhurst College in Illinois. During World War II he was an instructor for the Air Corps Technical School and served as a weather officer in Thule, Greenland. After leaving MC in 1965, he finished his teaching career at North Park College in Chicago and retired in 1987.

He was indeed my mentor and role model and I suppose it is easiest to reflect on Baumgart by thinking of one of his typical classes. I enjoyed them so much that I always took the same seat on the aisle in the front—I didn't want to have my view blocked or to miss hearing something important. He always entered the room neatly dressed for every class and was always prepared, enthusiastic, and ready to teach. He had the unique skill of making mathematics seem easier than it really was. He always wanted to minimize anxiety about math and often made a reassuring comment, "This is quite easy" even though the topic may have been otherwise.

*John K. Baumgart.*

One of his distinctive teaching skills was to visually bring math to life. He used colored chalk in his lectures and brought models to use as visual aids in class. He seemed to relish finding ways to help students see and understand the subject better.

As his teaching assistant, I was privileged to learn numerous skills that I still use in teaching and writing. Of particular importance was the preparation of tests that can be read and understood by students, cover the subject matter adequately, and be accomplished in the allotted time. Another perhaps simplistic teaching tip was to return the graded tests the next class period. No math student wants to wait a week to know how he or she performed.

John K. Baumgart was a class act through his entire career. The following are some comments by others whose lives were enriched by knowing him.

Stan Beery, now the Chairman of Mathematics at Manchester, studied under Baumgart. Beery said,

> *He had a great sense of humor, which made class even more interesting. For example, he loved the use of puns. He also possessed a real artistic skill at making meaningful drawings on the blackboard. This was before the availability of transparencies and computer-generated drawings. He was a master at the use of colored chalk. He presented topics clearly and crisply. I enjoyed his calculus class as much as any class of any kind I ever took. He was just a great lecturer.*

Beery recalls hearing statements like the following from other Manchester students who studied under Baumgart: "I never enjoyed math, but after studying with Professor Baumgart, I did love the subject."

Lois (Kiefaber) Peterson, Manchester graduate, and present professor of physics at Whitworth College, says:

> *Professor Baumgart gave me my first B at MC. I was impressed by his high standards. Rather than feeling resentment, students respected him for those high standards. He had a delightful sense of humor. I remember the smiley face to the zero power to emphasize that any nonzero number to the zero power is 1, and that the integral of 1 divided by d-Cabin is log Cabin + C, or "houseboat." There was also a gentleness and kindness about him that came through to students. He never put students down. He demanded respect from his students, but in exchange he was kind and respectful to them as well.*

Baumgart embellished my regard for mathematics textbooks. He was proud of the texts he used and spoke of them with high regard. His reverence of them inspired my interest in studying from them and exploring why they were good.

One such book was *Elementary Differential Equations* by Earl D. Rainville. Differential equations contains an elegant, well-structured theory using elementary calculus to find solutions, as functions, of equations involving derivatives. Baumgart loved this book because the exercises were truly fun to do.

REFERENCE

Rainville, E. D. *Elementary Differential Equations*, 2nd ed. New York, Macmillan, 1958, p. 141.

Baumgart enjoyed bringing humor to the class whenever he could. He took great pleasure in pointing out to us what Rainville wrote in this Exercise 32:

**32.** Solve "$\ddot{x}^2 + 4\dot{x} + 5x = 8\sin t$; when $t = 0$, $x = 0$, and $\dot{x} = 0$. Note that

$\dot{x} = \dfrac{dx}{dt}$, $\ddot{x} = \dfrac{d^2x}{dt^2}$ is a common notation for derivatives with respect to time

when the independent variable is time.*

. . .

*The disadvantage of such a notation when derivatives of higher order appear must be evident. Other objections can be raised; *flies have been known to perform undesired differentiations.*"

It was typical of Baumgart to draw a smiley face on the blackboard or on tests he graded. I still have my copy of Rainville's book, and the smiley face still peers up at me on that page whenever I open it. I also have a fifth edition of the book, revised after Rainville's death by Phillip E. Bedient, and it is a bit sad to note that the humor described above does not appear in that edition. It boils down to an adage I have developed as an author: What you might do very successfully in class may fall flat on the printed page.

An example of Baumgart's subtle humor is the following holiday handout.

## Holiday Wish

Do the following exercises and enter your answers in the spaces below.

**1.** Find $\displaystyle\int_0^M dx$ .

**2.** What is the value of $\displaystyle\lim_{x \to 0} (1 + x)^{\frac{1}{x}}$?

**3.** Given: $y = u^3 + 2$ where $u = \dfrac{x}{3} + 5$,

Find: $\dfrac{dy}{dx}$ when $u = r$ .

**4.** If $x = \dfrac{1}{2}y^2$ find $\dfrac{dx}{dy}$ .

**5.** Find $\dfrac{dy}{dX}$ for $X^2 - 2y + 5 = 0$ .

**6.** If $y = \dfrac{x^3 + m^3}{x + m}$ find $\dfrac{dy}{dx}$ when $x = 0$.

Hint: Simplify before differentiating.

**7.** Solve for $x$ if $\log 10^x = a$.

**8.** Find the slope of $y = x^3 + 4x^2 + sx + 5$ at $x = 0$.

1    2    3    4        5    6    7    8

In addition to a bit of humor, the following story regarding Baumgart involves an element of pathos. The issue of cheating can be painful to an instructor. In a legal sense, it can boil down to an issue of the instructor's word versus the student's. Baumgart often prepared his exams in a multiple-choice format with small boxes for the student to darken correct answers. Baumgart virtually never prepared dual forms of a test. In those days, with no word processor, it was a time-demanding job.

On one of those multiple-choice tests Baumgart noticed a weak student sitting next to a top student and copying the darkened answer boxes. On the next test, he deliberately prepared two forms and gave form A to all the students except the cheater, to whom he discreetly handed form B. Baumgart often graded exams as each student handed them in. The cheater handed his in, and of course, the darkened boxes on his form B was in one-to-one correspondence to form A. Baumgart went through the exam, marked each answer wrong, turned to the front, and wrote score 0, grade F at the top. The guilt of the cheater shone in his redfaced expression as he looked at Baumgart, stating in unspoken words, "But I have the same answers as Tania!" No further proof was needed!

Baumgart embraced conversation. I recall many times stopping in his office to ask a math question, only to have our meeting evolve into a discussion of life and philosophy. I will never forget our discussion about the existence of God when he said that you could analyze the issue of the existence of God ad infinitum, but in the final analysis it boils down to the acceptance of an axiom. Being a math major, that concept has endured in my mind and conversation the rest of my life. It is a "faith axiom" of mine that God exists. This is indeed the foundation of my premise that the philosophy of every mathematician should embrace a union of faith and reason.

It was near the time of his retirement that I found myself searching for someone to produce videotapes to accompany one of my textbooks. I considered, "Who is the best lecturer I ever had?" The answer was John K. Baumgart. He wrote and lectured in those tapes with his usual outstanding care and quality. John K. Baumgart served Manchester College as associate professor of mathematics from 1956 to 1965. He passed away in February 1994. I miss him, but those who demonstrate mathematics teaching of the utmost quality serve as a constant reminder of this unique human being.

## DAVID L. NEUHOUSER

*David L. Neuhouser.*

David L. Neuhouser was an undergraduate math major at Manchester College and received his Ph.D. in mathematics education from Florida State University. He taught at Manchester College from 1959 to 1971, then moved to Taylor University, where he taught until his recent retirement. He received many awards for teaching excellence including the Sears Roebuck Foundation Teaching Excellence and Campus Leadership Award in 1989.

Neuhouser was my first college professor at Manchester College in the fall of 1959 when I took calculus as a freshman. That class was also the first college math class he ever taught. His teaching was inspirational and his calculus class was a turning point in my mathematical self-esteem in that I began to realize my talents and passion for math. His teaching conveyed a humble, kind, patient quality. He is the most humble man I ever met. Disappointingly, after I had studied calculus under him my freshman year, I had only one more course with him when I was a senior, secondary math methods. The rest of my courses were under Baumgart.

It is ironic that I virtually lost touch with Neuhouser over the years, though he was less than 60 miles away at Taylor University. In more recent years, however, he has touched my life from another side of his intellectual passion, through his writings connecting faith and reason.

Neuhouser developed a deep interest in Abraham Lincoln, C. S. Lewis, and George MacDonald, a theologian and author who was a mentor of Lewis. At Taylor University Neuhouser formed the C. S. Lewis Society, which meets five times per year and has a C. S. Lewis library collection. He received the Howard Vollum Writing Award for his article "C. S. Lewis and Mathematics," which appeared in *Faculty Dialogue* in 1995. His recent book, *Open to Reason,* is a comparison of the roles of reason, experience, imagination, intuition, faith, love, humility, and obedience in mathematics and religion. Neuhouser's interest in mathematics and religion embodies my strong belief that mathematicians and scientists need to embrace both their reason and their faith. Indeed, the next-to-last chapter of his book is entitled "Paradoxes, Postulates, and Pacifism," in which he lists the postulates, or axioms, of his Christian faith. More books like this need to be written by mathematicians and other kinds of scientists.

## GRADUATE SCHOOL PROFESSORS

I took the Graduate Record Exam (GRE), in mathematics in the fall of my senior year at Manchester and earned a fine score of 91. I was arrogant enough, or stupid enough, to think I could do better in the spring and took the exam again, actually to make Professor Baumgart look good. I remember

REFERENCES

Neuhouser, D. L., "C. S. Lewis, George MacDonald, and Mathematics." Elgin, IL, http://www.tayloru.edu/upland/programs/lewis/articles/article.html.

Neuhouser, D. L., *George MacDonald: Selections from His Greatest Works*. Elgin, IL, Scripture Press, USA, 1990.

Neuhouser, D. L., "C. S. Lewis and Mathematics." http://www.tayloru.edu/upland/programs/lewis/articles/csl_math.html.

Neuhouser, D. L., *Open to Reason*. Upland, IN, Taylor University Press, 2001.

him handing that test to me saying, "It looks easy!" Well it wasn't—it was an entirely new test and I got a 73 on it.

In hindsight, I did a very poor job of investigating graduate schools, applying to Ohio State University (OSU) because I was from Ohio and to Purdue because I had a friend who studied there. I was accepted at Ohio State but rejected for a graduate assistantship from Purdue because of my 91 then 73 on the GRE. Two years later, however, I received an assistantship at Purdue for my doctorate in math education.

In any event, coming from a small campus like Manchester (1,100 students) with a truly caring environment to a large campus like OSU (over 50,000 students) was a real jolt to my psyche. I should have applied to a smaller university like Bowling Green, Miami University of Ohio, or Colorado State, each of which offered no more than a master's degree in mathematics. Such an experience would have struck a better balance between a small college dedicated to teaching and the large, publish-or-perish math department at OSU. I found good teachers at OSU, but the road was rocky.

I have a great interest in the notion of paradox and my personal paradox began when I graduated from tiny Manchester College and entered a huge university like Ohio State. Talk about a stranger in a strange land. I was taking graduate courses way over my head, and 200 miles from Elaine, whom I missed desperately even though we were not married at the time. She graduated from Manchester during the first year of my master's program and began teaching my second year. It was indeed the loneliest and most difficult time of my life.

In hindsight, I know that I should have insisted on being placed in lower-level graduate courses with a more basic proof orientation at OSU. Although I had had sound math courses at MC, the lack of emphasis on proof there was a major problem when I faced my first courses. Ohio State was attempting to expand its Ph.D. program at the time and to that end accepted about fifty new graduate students into the master's degree program. What did the powers-that-be do with these students to foster their success? They threw us to the wolves with two of the worst professors in the department, then topped it off by having no textbook for either course. Consequently, I received two Cs in my first quarter at OSU, and was placed on probation. (Keep in mind that this was 40 years ago. I have no knowledge of the present math department at OSU.)

To the credit of the department, it took the C students, and there were many of us, and put them into two recovery courses the next quarter. By God's grace and the teaching of Angelo Margaris, I received an A and a B in each of the next two semesters, and pulled myself back to a 3.00 GPA and off probation. (Incidentally, the department also cut my graduate assistantship back from $240/month to $150/month. I was barely surviving financially.) I credit Angelo Margaris, for the most part, in salvaging my opportunity to earn a master's degree. His sound teaching of general topology, out of a book by George Simmons, complimented with a strong dose of proof techniques, became a mathematics survival handbook for me.

## Angelo Margaris

Angelo Margaris earned a Ph.D. in electrical engineering from Cornell University. After teaching at Oberlin College, he went to OSU and was an associate professor of mathematics when I studied for my master's degree in mathematics from 1963 to 1965. He later moved on to Southwestern College, now called Rhodes College, in Memphis, Tennessee. He published articles in the *Journal of Symbolic Logic* and the *American Mathematical Monthly.* His most important contribution was a book, *First Order Mathematical Logic*, published by Blaisdell Publishing Company in 1967 and later republished by Dover Publications.

Earlier I mentioned the notion of paradox, or enigma, in my own situation and I must say that also applied to Margaris. While his teaching techniques were outstanding, he had an intimidating side to his personality that could cut a student in two. You might stop him in the middle of a lecture to ask a question, and he would respond with an irascible statement like, "What's the matter with you; I just explained that!" Yet, the next time you asked, he would give you a nice explanation and compliment you on your question. Because you never knew his reactions, you sat in class in fear, yet paradoxically confident that he was an excellent teacher. Margaris smoked Camel cigarettes in class, and more often than not one was dangling out of his mouth, adding to the intimidation.

I have usually been able to see the other side of people who may seem tough on the exterior but have a kind heart on the interior. Margaris may have been my greatest test of this tenet, yet I had that faith in him when other students may have been scared to death. I might have been fearful as well, but I knew I was learning mathematics.

I have two favorite Margaris stories. We were going over our proof homework in class, and students were invited to give their proofs at the board. A brilliant student might create a short, elegant proof of a result, while I was delighted to just come up with a proof no matter how many pages it took. That was the situation one day when I volunteered to give my lengthy proof. I was at the blackboard working in detail through my proof when Margaris abruptly interrupted saying, "Where in the world are you going with this?" In my frustration, I snapped back at him saying, "Just be patient!" He allowed me to finish. My proof, though lengthy, was correct and accepted respectfully by Margaris. In the lounge after class, my fellow students could not believe I spoke out in that manner. Nor can I, to this day, believe that I had the audacity to defend myself.

The second story involves the book, *First Order Mathematical Logic*, which he was writing when I took another course from him on logic. He was class testing the manuscript with us, and each day he would come in and pass out newly written pages of his manuscript. The pages were not collated so he started at the front of the room and handed out the pile of p. 38, then the pile of p. 37, and so on. Now you have to keep in mind his sometimes gruff personality to enjoy the story. Imagine Margaris at the front of the room, with the ever-present Camel cigarette lit and dangling out of his mouth, informing us that "there is no p. 34." While he was passing out the pages, he repeats "there is no p. 34." After the stacks of pages were gone, he moved to the back of the room to collect the extra sheets, cigarette still dangling, and a third time announced that, "There is no p. 34." We could almost sense what was coming next. A student in the front of the room raised his hand and asked, "Sir, do you know that there is no p. 34?" You would have thought Mt. Vesuvius had erupted. Margaris stomped to the front of the room, threw the remaining pages up in the air, Camel dangling even more dangerously and said, "Oh my God, I told him three times that there was no p. 34!" As the pages fluttered to the floor, the rest of us wondered about the time of that student's funeral.

On the other side of the Margaris paradox, I was convinced that he had a good heart—so much so that I asked him to be my advisor for my master's thesis. At that time, a student had to take either a master's exam or write a thesis. I chose the thesis because I harbored a deep-seated fear of the test. Margaris was very helpful to me and the thesis went very well because I enjoyed the research and especially the writing. My fellow students marveled that I had chosen Margaris, but I took pride in working with him.

I always hoped that I would be able to relate to Margaris on a personal basis as I had with my professors at Manchester. I even thought that after I left for Purdue, he might call me to see how I was doing, but that never happened. As I was writing this book, 38 years later, I decided to try to track him down, and lo and behold, he was still alive in Ithaca, New York. He was very gracious to me and consented to read what I wrote about him here. I really did like him and especially admired his mathematics teaching skills. He must have even stopped smoking those Camels to have reached the age of 82.

I am convinced that God has his hand on your life and touches you in ways that you may never know. In our conversation Margaris revealed some information, unknown to me for 38 years. I commented that I received two Cs in my first quarter at OSU and went on probation with about 20 other students. In truth, the department's first reaction was to kick all of us out of the program, seeing no hope that any of us could ever get a Ph.D. Margaris said he and another professor, Jesse Shapiro, went to bat for us and convinced the department to create the two recovery courses that salvaged many of our degrees. To quote Margaris, "Getting a Ph.D. in pure mathematics is so difficult! These people deserved another chance." At least two of us did go on to earn Ph.D.s, myself in mathematics education and at least one, Jane Babcock, in pure mathematics. Margaris touched my life more deeply than I had known. Where would my life be without his act of kindness? This revelation so touched me that at this writing I must attach an MPX to the experience.

**MPX** MUSIC: "I Stand in Awe," sung by People of Destiny Music, Track 16 of the CD, *America's 25 Favorite Praise & Worship Choruses, Volume 3*, recorded by Brentwood Music, Inc., One Maryland Farms, Suite 206, Brentwood, TN, 1995.

### Norman Levine

There are few really outstanding teachers, so when I found one, I cherished the experience. Norman Levine ranks near the very top of my list of all-time great lecturers. I took a summer-school course from him on general topology. We did not have a book but in his case, none was necessary; when he lectured, the mathematics just flowed. His words were clear, his handwriting was impeccable, and I understood everything he said. His teaching flowed into my brain like pouring math out of a gravy boat. That was my best course at OSU—I received an A and was at the top of the class.

It was unfortunate that I could never connect with Levine to take another class, but I will never forget him! When he passed away in 1983, the OSU Trustees said, "He prepared meticulously both for his classroom teaching and for his coordination of large enrollment courses." Although I never had the opportunity to establish a warm relationship with him, other than through my classroom performance, that man had a profound effect on my career as a math educator. He could teach!

### Robert A. Oesterle

After earning my master's in mathematics at Ohio State, I went to Purdue University in 1965 to work on my Ph.D. in mathematics education. My major professor was Robert A. Oesterle, a professor of mathematics education who had published several K–12 textbooks for the American Book Company. My story about him is brief, one moment that I will remember the rest of my life.

I was in Dr. Oesterle's office being counseled about my plan of study and I recall telling him that my main interest in math education was at the college level. Normally, a student in this field of study intends to do K–12 math education research and teach methods courses. Since neither was my intent, I remember his trepidation at my goal, but he was willing to accept me as his student.

As we left his office together and walked down the hall, he spoke to emphasize his willingness to work with me, encouraging me and even putting his hand on my shoulder. I had not sensed that much caring by any professor since I had left Manchester College two years earlier. That slight gesture gave me more encouragement than I can ever express. It is amazing how a touch on a student's shoulder can carry so much meaning.

My work with Oesterle took a turn for the worse in the next year as he became manic depressive. He would answer my questions in a very curt manner. As opposed to the encouragement I had earlier received, I was jolted into a world of virtually no conversation at all, and certainly no encouragement. I was devastated! The only help I received on my dissertation was occasional conversations with the other math educators, Robert B. Kane, Mervin L. Keedy, and Grayson Wheatley. If ever a person did a dissertation on his own, it was me.

Oesterle was somehow able to work with me enough that I did finish my Ph.D. in the summer of 1968. Unfortunately, two years after I left Purdue, Oesterle took his own life.

### Christoph J. Neugebauer

In my second semester at Purdue, I took a course called Principles of Analysis II, which was a study of real analysis and measure theory. The professor was Christoph J. Neugebauer, truly a master teacher. What happened each day in that course is still a marvel to me.

Neugebauer walked in the room each day, cool as a cucumber, wearing the same outfit—a sport coat over a blue button-down shirt. He brought nothing with him, no book, no notes, no nothing! He would take his coat off, hang it over the chair, pick up a piece of chalk, and begin the most beautifully choreographed lecture you could ever ask for. His ability to communicate at the blackboard was very much like Levine's, but Neugebauer had NO notes! Keep in mind that this was second-semester graduate analysis, not ninth-grade algebra.

After that course in the fall of 1965, I had no contact with Neugebauer until in 2002 I tried to call him on the chance he might still be around. Not only was he still alive, aged 75, but he was still giving those lectures in the same way on that course. Perhaps I might be able to pick a topic like quadratic equations and wing it through a lecture. But that lecture would suffer if I hadn't decided what equations I might sequence, such as those proceeding from integer solutions, to fractional, to radicals, and to those with complex solutions.

Just thinking about Dr. Neugebauer walking into that classroom cold and doing that day after day is beyond my mind's comprehension. When I asked him about it, he said that he knows the material so well that he can lecture on it in much the same way that a seventh-grade teacher might teach operations on fractions.

Neugebauer was born in Germany and served in World War II as a "gofer" boy in an airplane hanger of the Luftwaffe. After the war, he came to the United States as a teenager, studied undergraduate math at the University of Dayton, and received his Ph.D. at Ohio State. It is a distinct pleasure to honor such a consummate lecturer at this point in my life.

## FELLOW STUDENTS

When I was a college student from 1959 to 1968, the notion of "self-esteem" was not evident in our society. Nevertheless, two fellow students played a significant role in building my self-esteem and contributed to my emotional and academic survival during that time.

### Nelson L. Zinsmeister

Like every college freshman, I took English classes at Manchester College and, while I did not particularly excel, my instructor saw enough writing skill in me to recommend my name to the college newspaper. The editor of the paper, Nelson L. Zinsmeister, was a junior whose encouragement

*Nelson L. Zinsmeister.*

inspired me to want to please him. Although he assigned me to various writing projects, the most significant was a personal story about the maintenance people at the college. Nelson was so impressed with my article that he awarded me a by-line—my name was attached to the story! While that story did not qualify me to be a textbook author, it was a big boost to my self-esteem at a critical time. Me, be an author? No way, I just got B's in English.

Nelson was raised on a farm in nearby Bippus, Indiana (most famous as the childhood home of noted sports announcer, Chris Schenkel). He often took me home with him on weekends for some special meals and visits with his loving family. I learned that Nelson excelled in high school math under a brilliant teacher, Jim Rowe, and won some college scholarships. In contrast, although I was a good high school math student, no one offered me any scholarships, and I certainly didn't win any Putnam math contests. I was a late bloomer who watched too much TV in high school; my math grades were better in college.

Nelson was also a math major who wanted to be a secondary teacher. He and I struck a friendship that resulted in our becoming roommates during my sophomore, his senior, year. To this day, I wonder what Nelson saw in me to invite me to room with him. Maybe that question is what remains of my memories of a poor self-image at the time. In any event, he was a real hero to me. I found myself observing his math study habits, his way with people, and his general outgoing personality, and I realized that I could not go wrong to pattern my life after Nelson's.

I will never forget the A I got in the first quarter of advanced calculus as a sophomore. That course was tough and to Nelson, I was to be compared to brilliant mathematician David Hilbert. I believe it was his enthusiasm, not the A, that fired me up to be a math major and an educator. Nelson earned a 93 on his GRE his senior year and that quickly earned him a graduate assistantship at Purdue University. I hated to see him graduate, but we stayed in touch, and I remember how much I enjoyed visiting with him at Purdue.

After Nelson graduated, he married Mary Martha Miller and boosted my self-esteem to a new high when he asked me to be his best man. I was deeply honored, especially since Nelson had two brothers.

Nelson and Mary Martha moved to upstate New York and became outstanding high school teachers. I was told Nelson's students adored him, which came as no surprise.

In the fall of 1963 I was working on my master's at OSU and would commute back to Logansport to see Elaine every other weekend. It was at Elaine's home one Thanksgiving morning that I heard the tragic news announced over the radio: Nelson, Mary Martha, and her sister, back in Indiana at the time visiting relatives, were killed by a drunk driver running a stop sign near Markle, Indiana. It was believed that Mary Martha was pregnant. Elaine and I attended the funeral in Wabash on that Saturday and I was a pallbearer. That was the first deep, tragic loss in my life.

So many times over the years I have wondered about the impact Nelson would have had on his students and society. I do feel that as my textbook writing career expanded, I surely would have tried to inspire Nelson to join me as a co-author. My instincts are that he would have been an outstanding author, and that we would have thrived working together. But, only God knows.

In 1970 I dedicated my second book, *Logic and Proof*, to

> *Nelson L. Zinsmeister, I lost a wonderful friend*
> *. . . the world lost a wonderful math teacher.*

It is my Christian faith that God usually does not plan tragedies, but He does have the divine capacity to redeem a tragedy and resolve the paradox. If my life and writing career have in some small part redeemed the tragedy of Nelson's untimely death and honored his brief life, then God has blessed me.

I can't wait to see Nelson again in heaven! I'd like to know how many of his kindnesses to me were a part of a plot between God and him to build my self-esteem and encourage me to pursue being an author.

### James S. Biddle

I have expressed how overwhelming it was to attend huge Ohio State after being an undergraduate in the small, caring environment of Manchester College. But, again, God has his ways.

One of my office mates at that time was James S. "Jim" Biddle, who came to Ohio State at the same time I did, but with a master's degree from Bowling Green State University. Needless to say, his math maturity was much higher than mine. Our friendship grew as we took classes together and shared teaching ideas and exams in our teaching assignments.

Jim was married to a sweet lady, Judie, and the two of them took me under their wing. Jim took time from his studies to tutor me in my math deficiencies, and on weekends when I was not away from Columbus seeing Elaine, he and Judie would have me over to watch Cleveland Browns football games and have supper with them. I will never forget their kindness to me when I was so very lonely and struggling academically.

Jim received his Ph.D. in abstract algebra and went on to become Dean and Director of the Lima Campus of Ohio State and founding President of Lima Technical College. He recently retired from Urbana College in Urbana, Ohio, where he was department chairman.

Jim and I lost touch for a number of years. We reconnected after I included my e-mail address in one edition of my algebra-trig series. I lived to regret that decision because I got so many e-mails from students wanting me to do their homework and asking about how to get a student solutions manual or videotapes, issues they should have solved through their instructors. The high point of that decision, however, was receiving Jim's e-mail. We have gotten together a few times since then and now keep in touch via e-mail. He has reviewed this manuscript, for which I am deeply appreciative.

*James S. Biddle.*

Sadly, pain and suffering has touched the lives of Jim and Judie Biddle as well. They have two daughters, both of whom are inflicted with a tragic genetic disorder called spastic paraplegia, a progressive disorder for which there is no cure and little treatment. The girls still live at home and require full-time care, so Jim and Judie are somewhat confined because of their dedication to their daughters.

I yearn to solve their problems, but it is not in my power. It is hard for me to see how God might resolve the pain and suffering in their lives. Even though their pain may not be resolved in this lifetime, I always sense their great, positive attitude. I will always treasure the friendship and kindness of Jim and Judie Biddle.

## THE UNITED STATES AIR FORCE ACADEMY

This is a story of flat-out joy! As a child, I dreamed of being a pilot. Not only did I spend hours building and flying model airplanes, but I read extensively and enjoyed movies about flying. Unfortunately, the dream of being a pilot was burst by the same physical complication that affected my playing baseball—poor vision.

*The United States Air Force Academy.*

But God gave that dream back to me in 1978 when I became a Distinguished Visiting Professor at The United States Air Force Academy (USAFA). The fulfillment of that dream started with the writing of our paperback book *Algebra and Trigonometry: A Functions Approach* in 1974.

I was attending a convention when two Air Force officers approached me in the Addison-Wesley booth. Colonel Paul Ruud and Colonel Tony Johnson were math AFA instructors, who informed me that they were using our algebra-trig book for cadets who needed extra instruction before starting the calculus sequence. They were very complimentary of the book and invited me to visit the math department at the academy. During a subsequent visit they offered me a one-year position as Distinguished Visiting Professor. It was typical at that time for each USAFA department to have such a position for a civilian professor. I was honored but could not get away for a full year. I did volunteer for a two-month position in 1978 without pay.

I can only describe that experience as walking into the "Garden of Eden" of education. Each small classroom was lined with blackboard space so the cadets could work at the boards and instructors could move around to help them. Each time an officer-instructor walked into the classroom, the cadets would stand and salute. In my case, they stood without saluting. It was quite a compliment and indeed a refreshing change from the academic climate in a civilian university.

As an instructor, I have always felt that if I could ever get the proper study time from my students, I could give them success. Analogous to "Location, Location, Location!' in the real estate business, my motto is, "The three most important skills in mathematics learning are Time! Time! Time!" I competed for my students' time. In a civilian university, an instructor competes for time from the demands of work, family, leisure, and other courses. At USAFA, the competition comes primarily from other courses. If there was a physics exam on a particular day, you could pretty much count on your cadets not having done their math homework. On the other hand, if there was a math exam, you almost had their undivided attention.

As opposed to a civilian university, when we weren't teaching, we were expected to be in our offices for tutoring or consulting with the cadets in our classes. I thrived on this guideline. When I was not tending to my cadets, I was either preparing for my classes or writing. Even though my time was short, I was able to build some nice relationships with a few of my cadets.

The strongest relationships I made were with some of the officers in the department. Colonel Tony Johnson was a class act as a friend, an officer, and an instructor. Colonel Paul Ruud, Major John Field, Major Jim Higham, and Lieutenant Colonel Robert Lochry, the chairman of the department, were several others with whom I built friendships. There was such a feeling of camaraderie and patriotism among the officers and me that I developed an almost irresistible desire to wear blue shirts and dark blue ties and suits.

Soon after arriving in Colorado Springs, I noticed the Air Force had a lingo and symbolism all its own. Usually, it involves letters strung together such as MOS, APB, and TDY, the latter meaning "Temporary Tour of Duty." There were so many acronyms that I couldn't help but write a poem using all the lingo, and the department officers got a big kick out of hearing that in the seminar.

*T-37 jet trainer.*

The most memorable event of my two months at USAFA was the thrill of a guest flight aboard a T-37 "Tweety-Bird," a jet trainer like the one pictured at the left. Major Jim Higham took me on that ride. I had to borrow a flight uniform from another officer, and that was just the first thrill. Jim drove me out to the field, where I had to go through some preflight training that almost scared me out of the experience. First, a training officer pointed out the location in my uniform of the snake-bite medicine. My horrified response was "The snake-bite medicine! What in the world do we need that for? Are there snakes in the cockpit?" The training officer then told me I might need it if I had to bail out of the plane and landed in a forest. "Bail out! Bail out! Not me!"

The next preflight training experience was learning how to vacate the plane in the event of a fire on the runway. I had to climb out of the pilot's seat and slide down the body of the plane to the ground. The thought of fire on the runway intensified my fears.

The crowning blow came when I climbed into a simulated cockpit and was taught how to fire the ejection seat. To eject, I would have to press my legs together, grip the firing mechanism handles on the sides of the seat, and pull. There followed a lesson on what to do if my parachute failed to open.

That did it—I was ready for some Pepto-Bismol and the nearest couch. But I was determined not to let this experience pass.

Much to my dismay, there was more to come. We finished the preflight training and went out and climbed in the plane, and I thought to myself, "You are going to be a fighter pilot." Then Jim said to me, "I have one more warning. Most of the cadets I take up on these flights think they are real flight jockeys. They ask to do loops, wingovers, and other stunts. Then many of them get sick to their stomachs and throw up in the cockpit. If you feel like you are getting sick, please tell me immediately!" I agreed, and off we went.

We taxied down to the end of the runway and Jim checked with the tower. They approved his takeoff but told him to make a sharply banked left turn after takeoff to avoid some bad weather that was approaching the end of the runway. I have a nervous stomach and have never thrived on roller coasters or wild rides. When we left the ground and banked to the left, I started turning green from pulling G forces. My first thought was, "If I tell him I'm sick, he will land right away." But I knew I should tell him, and I did. He leveled the plane and got some cool air into my flight mask, but even then it remained tough to breathe through. From there the flight went very smoothly. My only regret was that I did not let Jim do an aileron role, a gradual rotation of the plane around an axis through the middle of its body, but I was just too afraid I would get sick.

A few days later, I got to take a glider ride, and that was a world of difference. It is so pleasurable to fly and not hear engines roaring—much like a bird I suppose.

With the people in the department, the cadets, the flight experiences, and the beautiful scenery of Colorado Springs and Pike's Peak, that two months was a highlight of my life. When I left two months later, I had tears in my eyes. I hated to leave.

Some years later at a math convention, I noticed a booth in the book exhibit area for the Navy. Being curious, I inquired about its goal for being there. The sailors said they were hiring math instructors to serve aboard aircraft carriers. In case you haven't deduced it by now, I have an adventurous streak—and did that idea ever appeal to me! I could be below deck and teach algebra-trig, and then go up on deck and watch those F-14 Tomcat jet fighters take off and land. I might even catch another ride with a Top Gun pilot! Unfortunately, to do so, I would have had to be totally away from my family for six months, and I just could not impose such an absence on them. Had I been single, I would have signed up right then and there.

# On Writing Mathematics Textbooks

# 6

REFERENCE

Kosslyn, S. M., and Rosenberg, R. S., *Psychology: The Brain, the Person, the World.* Boston, Allyn & Bacon, 2001, pp. 302–309.

---

*A story may do what no theorem can do. It may not be (like real life) in the superficial sense: but it sets before us an* image *of what reality may well be like in some more central region.*

—C. S. LEWIS, *Of This and Other Worlds.* London: William Collins Sons, 1980, p. 39.

REFERENCES

Bittinger, M., *Introductory Algebra,* 9th ed., Boston, Pearson/Addison-Wesley, 2003.

Bittinger, M., *Basic Mathematics,* 9th ed., Boston, Pearson/Addison-Wesley, 2003.

Bittinger, M., *Intermediate Algebra,* 9th ed., Boston, Pearson/Addison-Wesley, 2003.

## INTRODUCTION

Psychologist Stephen M. Kosslyn of Harvard University defines creativity as, "the ability to produce something original of high quality or to devise effective new ways of solving a problem." Can the creative aspects of writing be taught? Psychologists tell us that creativity is one of the hardest topics to understand and to teach. It is my conviction that it is a gift, indeed for me a spiritual one. Thus, it may be an impossible task to teach authorial creativity to my reader, especially as a list or recipe of steps to success. Nevertheless, I am going to try to convey as many aspects of creativity as I can, in the context of story telling. Thus, I will try to convey what it is like to be an author by doing what C. S. Lewis calls, "setting before you an image of what it is like."

## THE BIRTH OF THE TRILOGY

I started the story of the birth of the Trilogy in Chapter 1. The Trilogy consists of three paperback books on developmental mathematics, *Basic Mathematics*, *Introductory Algebra*, and *Intermediate Algebra*, now in their ninth editions.

As a student, I admired my math books like some people admire artwork. The symbolism and the graphics on the pages gave me as warm a feeling as did being on a finely groomed baseball diamond or walking amid the splendor of my beloved Canyonlands of Utah. I wondered in awe how anyone could ever write a math book.

The title of my first book, co-authored with Mike Keedy, was *Trigonometry: A Programmed Approach*, published by Holt, Rinehart & Winston in 1969. You may not be familiar with programmed learning, a process by which very brief frames of writing or learning concepts are presented, followed immediately by questions for reinforcement. Here are the first few lines of the text.

*SETS*

*We shall primarily be dealing with sets of real numbers. For example, the integers 1, 2, 3, 4 form a set. We shall denote this set {1, 2, 3, 4}. Henceforth, we shall use the braces { }, to denote a set, the members of the set being named between the braces.*

> **(a) The set that contains the numbers $\frac{1}{2}$, 2, $-1$ is denoted**
> _____.

> **(b) The members of the set $\{1, \pi, \text{Charlie Brown}\}$ are denoted**
> _____.

*(a)* $\{\frac{1}{2}, 2, -1\}$

*(b)* $1, \pi$, Charlie Brown

Programmed learning was an educational fad for a short time in the late 60s and early 70s, but it soon faded out because the learning "chunks" were too short and not conceptual enough to build skills meaningfully. Nevertheless, that project was a great training ground for my future writing because

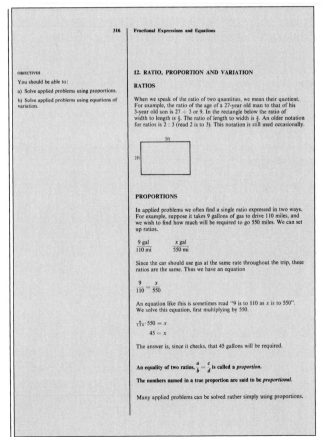

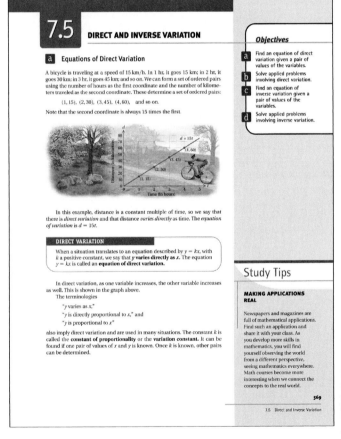

FIGURE 1

*A page from the first edition (left) and a page from the ninth edition (right) of* Introductory Algebra, *published in 1971 and 2003, respectively.*

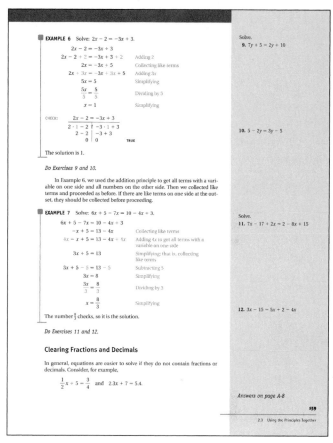

FIGURE 2

*A page from the first edition* (left) *and a page from the ninth edition* (right) *of* Introductory Algebra, *published in 1971 and 2003, respectively.*

not only did I learn pacing and flow, but I was forced to deal with the idea of behavioral objectives.

The editors at Addison-Wesley started us out with a page design that had a very wide margin. As authors, Mike and I were charged with figuring out how we could use the margins in an effective manner. Figures 1 and 2 show sample pages comparing the first edition in 1971 with the ninth edition in 2003. They reveal how the creative process evolved over thirty-two years based on feedback from students and instructors.

We tried different uses for the margins but the innovative use of the margin that evolved was adapted from our experience with programmed learning. The concept was to have a short stretch of textual material, which might be roughly one-half page long. Then the student would see a direction line like "Do Exercises 9 and 10." Students were to stop reading and do the directed exercises. By striking this middle ground between "brief-chunk" programming and the "long-chunk" setup of a normal text, we got students more consistently involved in the learning. This system went along with what math instructors always say, "The best way to learn math is with a pencil and paper in hand."

If you cover the margin of a page in our textbooks, what remains looks just like a regular textbook. So, with our books, not only did the students get more for their money, they got a *learning system* that enhanced the readability of the text. Students really could read and learn from a math book more effectively when it was *designed for learning*. This creative discovery was an **MPX**.

**MPX** MUSIC: "The Great Gate of Kiev," from *Pictures at an Exhibition*, Track 11 on the CD *The Fantastic Leopold Stokowski Transcriptions for Orchestra*, by Erich Kunzel and the Cincinnati Pops Orchestra, 1994 TELARC.

But there is more to the story of these books. When I was studying for my Ph.D. in math education, there was considerable interest in a little paperback called *The Taxonomy of Educational Objectives* by Benjamin S. Bloom. Until then, no one had written a book listing the behavioral objectives for the students. The upper right-hand margin of the opening section of the text lists behavioral objectives coded with letters [a], [b], [c], and so on.

Mike Keedy brought the idea of using objectives to our books. Not only did the objectives let students know exactly what skill was to be learned before moving to the next section, they also forced us as authors to structure our writing so that we taught the objectives in a linear manner. We taught [a], then [b], then [c]. Simple as this may seem, no previous authors had had the courage or the insight to write using objectives. These coded objectives appeared in the text, then in the exercise sets, the review exercises, the chapter tests, and the cumulative reviews. By following the codes of section number and objective number, which were placed next to exercises or answers, students had a handy way to go back and review exercises they had missed or found difficult. For example, an exercise coded [2.3b] dealt with material in section 2.3 pertaining to objective b.

Another phenomenon was alleviated by listing objectives and then writing accordingly. It is what I call "the treasure hunt phenomenon." When students prepare for a test, they are usually not completely sure what to study, and instructors hope that by not telling exactly what the behavioral objectives are, they are encouraging the students to study everything that has been taught since the last test, as if on a treasure hunt of what to study. When I taught out of these books, I always told the students that I had no secrets. Indeed, the objectives are listed at the beginnings of sections, and there is a sample test at the end of each chapter. They never needed to ask me what could be on the test—they just had to look at the objectives and the sample tests. Establishing behavioral objectives eliminates the "treasure hunting" phenomena.

These innovative books tested the perseverance of the Addison-Wesley sales reps to overcome the resistance of instructors who were set in their ways. The system was explained and the benefits were presented over and

*I thank Addison-Wesley for persevering so long with these books. It was 8 to 10 years before they began to dominate their market. When a book does only so-so sales in an early edition, a publisher may stop revising it and let it go out of print. But AW stuck with us—again I thank them sincerely.*

over. Then students verified that the books worked, and colleges who used the books spread the word. Finally, in the third edition when a second color and an enhanced design were used, the books really took off and created their own paperback market—the first ever developmental math series in paperback. The AW rep in Arizona, Jack Sandbo, was so successful with the books that at one time *every* college in the state used at least one of these books. And it was very satisfying to build something we could hold up proudly as a finished product, and then have so many students learn from it successfully.

### Beyond the Trilogy

The snowball began rolling down the hill. As Addison-Wesley found out we could write books they could proudly sell and students could learn from, the editors approached us to write other books—algebra and trigonometry, brief calculus, precalculus, hardback algebras, and so on. In January of 2003, I finished writing the ninth edition of the Trilogy and the 165th math textbook in all, if you count the revisions. God has blessed me with this ability and I am truly thankful to Him.

## THE KEEDY–BITTINGER WRITING PHILOSOPHY

*Make everything as simple as possible, but not simpler.*
—ALBERT EINSTEIN,
1879–1955

Mervin L. "Mike" Keedy was a mentor's mentor. Being able to learn from one of the masters of textbook writing provided a real jump-start to my writing career, a tremendous opportunity and one that I embraced with all the passion I could muster. Once an envious colleague commented that "the only reason Bittinger was successful at writing was because of his connection to Mike Keedy." There's no doubt that Mike accelerated my success, but I feel to the bottom of my heart that I contributed hard work, dedication, creativity, and talent to our collaborative efforts. I dare say that without Keedy I still would have become a successful textbook author. But with him, I became one quicker and was more prolific because he was such a terrific mentor.

### Keedy and the Early Years

Keedy was an outstanding author in the modern math movement of the late 60s and early 70s. That movement had many shortcomings, the first being an emphasis on teaching higher-level mathematics at lower levels. It is possible to teach nondecimal bases to third graders, but is it worth the price? I doubt it! It is also possible to increase rigor in many kinds of secondary mathematics classes but, again, is it worth the price? You might be able to teach some aspects of calculus to ninth graders, but if you sacrifice skills at that level to do so, the lack of skills will haunt both students and their instructors at a later date.

One aspect of the modern math movement to which both Mike and I adhered in our writing was the idea of providing understanding, a rationale, to the mathematics we wrote about. But, we did it *without sacrificing skills.* Thirty years later, I still ascribe to that philosophy in writing every one of my books! I try to write toward understanding, but I want sound mathematics skills as a result.

### Discovery Learning

REFERENCE

Bittinger, M. L., "A Review of Discovery," *The Mathematics Teacher,* Vol. 61, 140–146, 1968.

Keedy, M., Jameson, R., and Johnson, P., *Exploring Modern Mathematics, Books 1 and 2.* New York, Holt, Rinehart, and Winston, 1968.

Keedy, M., Dwight, L., Nelson, C., Schluep, J., and Anderson, P., *Exploring Elementary Mathematics, Grades K–6.* New York, Holt, Rinehart, and Winston, 1970.

One way to provide a rationale behind mathematics concepts is by *discovery learning.* As an author and an instructor, I must warn that one cannot write a math book in which all concepts are discovered; it would be the size of a big-city phonebook. Similarly, if you attempt to have students discover every concept of a particular math course, you will not have enough classroom hours. But, with apt choices, the notion of discovery learning can enhance the writing of a book and your teaching.

I believe the two most creative math books ever written were for the seventh and eighth grades by Mike Keedy and his co-authors. They were a wonder of discovery learning and they should be examined for their excellent learning methods, especially their use of discovery. The books eventually became victims of the backlash against the modern math movement because of a lack of skill exercises and applied problems, but they were fine examples of the possibilities of creative teaching.

Because of the success of these junior high books, Mike was just finishing the publication of an elementary series when he and I signed with Addison-Wesley to write the developmental math series known as the Trilogy.

The first of these three books, *Basic Mathematics,* teaches the skills for seventh- and eighth-grade arithmetic. With his experience writing K–8 math books, Mike actually wrote the outline for this book one evening in his hotel room. That outline and the sequencing of topics that Mike had learned from those K–8 books were a superb basis for *Basic Mathematics.* Two obvious examples come to mind. The first was the placement of the topics of multiplication and division using fractional notation before addition and subtraction. The logic for this sequence was that multiplication and division with fractional notation is much easier to carry out and conceptualize than addition and subtraction are. But such issues are not always easy to get across to teachers. You see, tradition is a weighty factor in the minds of those who adopt textbooks. On the one hand, instructors look for new and innovative teaching materials, but on the other hand, they don't want the boat of their traditions rocked too much.

A second example of how we benefited from Mike's writing experience was the inclusion of the notion of multiplying by one. Look at the following example.

$$\frac{2}{3} = \frac{2}{3} \cdot 1 = \frac{2}{3} \cdot \frac{4}{4} = \frac{2 \cdot 4}{3 \cdot 4} = \frac{8}{12}.$$

How can we change a denominator in order to add using fractional notation? Our way is to multiply by "one" using an appropriate symbol for 1. Multiplying by 1 does not change the number; it only changes its symbolism. To us, this provided a stronger rationale for changing fractional symbolism than the magic trick, "multiply the numerator and denominator by the same number." Why does that work? Without the multiplying-by-one concept, it is difficult to provide a worthy rationale.

The notion of multiplying by 1 provides a rationale for simplifying fractional notation. We factor and reverse the process of the preceding equation. Doing so until there is no common factor other than 1 yields the simplest fractional notation.

If you examine all my books from *Basic Mathematics* through *Algebra-Trigonometry* to *Brief Calculus*, you will still see the notion of "multiplying by 1" used as a rationale for many skills and concepts; it is a unifying concept of these books.

## Canceling

When teachers have taught the same course for years and have their lesson plans and tests geared a certain way, it is hard to undo that tradition, even if a better way is obvious. An author must always write amidst the dynamic tension of "teaching tradition" vs "author preference and philosophy," which means we have backed off in some instances because of overwhelming teacher tradition. The issue of "canceling," a word I detest in my textbooks, is a such an example.

One of the notions of the modern math movement that Keedy and I chose to follow is never to mention "canceling" in our books, at least as a skill. Indeed, the word does not appear in the first edition of *Basic Mathematics*. It is common when a group of math instructors are together to hear them tell amusing war stories involving what Keedy called the "Student's Law of Universal Cancellation." The law asserts that anything above the line can be scratched out, or "canceled," if it has something to match it below the line. This leads to the following erroneous and amusing examples.

*You want an example of what mathematicians do for research? I once met a mathematician who did an entire research paper entitled "Cancelable Numbers," describing the characteristics of pairs of numbers that can be "canceled" in this manner to give correct results.*

*There are situations where this false procedure does work. For example,*

$$\frac{16}{64} = \frac{1}{4}.$$

a. $\dfrac{1\cancel{5}}{\cancel{5}4} = \dfrac{1}{4}$*   b. $\dfrac{\cancel{x}+1}{\cancel{x}+3} = \dfrac{1}{3}$   c. $\dfrac{\sin x}{\sin t} = \dfrac{x}{t}$

d. $\dfrac{\sin x}{n} = 6$   e. $\dfrac{x^{\cancel{2}}}{\cancel{2}} = x$   f. $\dfrac{x+2}{x} = 2$

To illustrate my point, if you are an instructor, write (e) and (f) in sequence on the blackboard. I will virtually assure you that there will be laughter in the room for (e), but very little, if any, for (f); though both are incorrect, students are just not quite sure. This is overwhelming evidence to me that the notion of "canceling" should disappear from textbooks and the classroom forever. I learned this little test from Keedy, but it became an `MPX` as I experienced this situation more than once as a professor.

`MPX` MUSIC: "Ferry 'Cross The Mersey" by Gerry and the Pacemakers, music and lyrics by Gerry Marsden. Dick James Music, Inc. BMI, Laurie 3284 (1965).

This relates to one of my early learning experiences as an author. In the first edition of *Basic Mathematics* we did not mention canceling. But much to my surprise, when reviews came in on the first edition, there was overwhelming demand from instructors to include canceling in our books. Much to our chagrin, it is now included, but with a carefully delineated warning like the one in Figure 3.

## CANCELING

Canceling is a shortcut that you may have used for removing a factor of 1 when working with fraction notation. With *great* concern, we mention it as a possibility for speeding up your work. Canceling may be done only when removing common factors in numerators and denominators. Each common factor allows us to remove a factor of 1 in a product.

Our concern is that canceling be done with care and understanding. In effect, slashes are used to indicate factors of 1 that have been removed. For instance, Example 6 might have been done faster as follows:

$$\frac{90}{84} = \frac{2 \cdot 3 \cdot 3 \cdot 5}{2 \cdot 2 \cdot 3 \cdot 7} \qquad \text{Factoring the numerator and the denominator}$$

$$= \frac{\cancel{2} \cdot \cancel{3} \cdot 3 \cdot 5}{2 \cdot \cancel{2} \cdot \cancel{3} \cdot 7} \qquad \begin{array}{l}\text{When a factor of 1 is noted,} \\[4pt] \text{it is "canceled" as shown: } \dfrac{2 \cdot 3}{2 \cdot 3} = 1.\end{array}$$

$$= \frac{3 \cdot 5}{2 \cdot 7} = \frac{15}{14}.$$

**CAUTION!**

The difficulty with canceling is that it is often applied incorrectly in situations like the following:

$$\frac{2 + 3}{\cancel{2}} = 3; \qquad \frac{\cancel{4} + 1}{\cancel{4} + 2} = \frac{1}{2}; \qquad \frac{1\cancel{5}}{\cancel{5}4} = \frac{1}{4}.$$

Wrong!        Wrong!        Wrong!

The correct answers are

$$\frac{2 + 3}{2} = \frac{5}{2}; \qquad \frac{4 + 1}{4 + 2} = \frac{5}{6}; \qquad \frac{15}{54} = \frac{5}{18}.$$

In each situation, the number canceled was not a factor of 1. Factors are parts of products. For example, in $2 \cdot 3$, 2 and 3 are factors, but in $2 + 3$, 2 and 3 are *not* factors. Canceling may not be done when sums or differences are in numerators or denominators, as shown here.

If you cannot factor, do not cancel! If in doubt, do not cancel!

FIGURE 3
*A page from* Basic Mathematics *showing strong cautions about using canceling.*
Bittinger, M., *Basic Mathematics, 9th ed. Boston, Pearson/Addison-Wesley, 2003, p. 136.*

## Improper Fractions

One of our beginning standards was to avoid unnecessary language. One example is the notion of so-called improper fractions, which we explain as follows:

*A fraction larger than 1, such as $\frac{27}{8}$, is sometimes referred to as an "improper" fraction. We will not use this terminology because notation such as $\frac{27}{8}$, $\frac{11}{9}$, and $\frac{89}{10}$ is quite "proper" and very common in algebra.*

Another such topic is the notion of naming an equation an *identity*, a *contradiction*, or a *conditional*. If an equation has no solution, it has no solution. Why does it need another word to describe and confound the student? If it has just one solution, or if all real numbers are solutions, so be it. For intermediate algebra students, this excess terminology just adds to their learning burdens. If you walked up to 1,000 students just after they've taken the final exam for an intermediate algebra course and asked them to describe a contradiction, how many do you think would answer correctly? The notion of an identity is useful in trig, but not before.

## "Keedyisms" and "Bittingerisms"

Our editors have gotten so used to talking to us about the many quirks in our writing that they often call them "Keedyisms" and "Bittingerisms." In every math book I have ever written the pronoun used is *we*. This came from Keedy. From time to time, our reviewers reacted negatively to this pronoun. To us, *we* is an informal word, connoting a relationship between the reader and the author. As an educator, I feel building a relationship with a student is an important aid to learning. But an author works in solitude, and I imagine students often think of us as unfeeling nerds who dish out exercises and word problems with no regard for the effort required by the student to learn. The use of *we* helps us to feel that the author and the student are working together in order to learn mathematics. *We* has been in our books for more than thirty years and it will stay another thirty as long as I make the decision.

## Writing to Learning Objectives

Keedy spotted two flaws in my writing style and insisted that I change them. The first occurred when there were two objectives to write about. Instead of covering the first and then the second, I would make the mistake of running the two together. The result was generally confusion. Mike often cautioned:

*If you have two objectives to write about, write about the first one, and then write about the second one. Don't mix them together.*

The second flaw in my writing was a common one, the result of spending most of the first twenty-six years of my life learning mathematics. My tendency was to put every math fact I knew into the writing. Unneeded verbiage confuses the student and adds to costs for both the publisher and the student. I learned to be aware always of length as I write! Less can be more.

## Class Testing a Book

When Keedy and I were writing our first book together, the programmed trig book, he insisted that we class test the manuscript. At that time copying was not easy, and I had to prepare the manuscript using green mimeograph masters. I underwent a hernia surgery at that time (my second of six; my stomach is made out of papier mâché) and recall working on those masters while I was in the hospital.

Over the years, I have had the pleasure of teaching out of my books several times. And I thank the math department at Indiana University–Purdue University at Indianapolis (IUPUI) for many such opportunities. A department could construe such class testing as an abuse of students or as offering less than the standard course. But I have never been denied the opportunity to teach from one of my books, even when it differed from the departmentally adopted book.

When teaching out of a book I had written, it was important to create a "safe" atmosphere with the students so they might offer constructive criticism. If I had class tested the materials believing that what we had written was perfect, I would have been wasting time and effort. Keedy encouraged me to tell students to write on the manuscript when they had trouble and to feel free to ask questions and make comments. Not only did I ask for their comments as I taught, I kept notes on the manuscript in order to rewrite it accordingly.

I have always learned from my students. Sometimes when they told me they couldn't understand what I had written, I tried something different or expanded into a general discussion right there in the classroom. Then if it seemed to be effective, I would rewrite accordingly. Likewise, when reviewers criticize a topic, I think about their points, then try something different in class. The more experience I gained, the more I could judge whether a new explanation would work.

One of the classic stories I tell involves the teaching of scientific notation. Suppose you want to convert 236,788 to scientific notation $2.36788 \times 10^5$. I remember, as a student, trying to learn rules about how you move the decimal point and then adjust the corresponding power of 10 accordingly when you convert. Do you move to the right, to the left, and does the power increase or decrease? I can still get confounded by those rules. We developed a more conceptual way to make those conversions, which is shown in Figure 4.

When teaching from my textbooks, I often had the good fortune to learn from my students, and I must sincerely thank and praise them for all their contributions to my writing. They deserve an MPX .

MPX MUSIC: "There! I've Said It Again," by Bobby Vinton, music and lyrics by Redd Evans and David A. Man, Music Sales Corp. ASCAP. Epic 9638 (1964).

No instance of such help was as evident as the following suggestion. When doing multiplication and division with scientific notation, one typically

You should try to make conversions to scientific notation mentally as much as possible. Here is a handy mental device.

> A positive exponent in scientific notation indicates a large number (greater than or equal to 10) and a negative exponent indicates a small number (between 0 and 1).

**EXAMPLES** Convert to scientific notation.

**15.** $78{,}000 = 7.8 \times 10^4$

$$7.8{,}000.$$
$$\qquad \text{4 places}$$

Large number, so the exponent is positive.

**16.** $0.0000057 = 5.7 \times 10^{-6}$

$$0.000005.7$$
$$\qquad \text{6 places}$$

Small number, so the exponent is negative.

Each of the following is *not* scientific notation.

$$\underline{12.46} \times 10^7 \qquad\qquad \underline{0.347} \times 10^{-5}$$
$$\quad\uparrow \qquad\qquad\qquad\qquad\quad\uparrow$$

This number is greater than 10.   This number is less than 1.

**EXAMPLES** Convert mentally to decimal notation.

**17.** $7.893 \times 10^5 = 789{,}300$

$$7.89300.$$
$$\qquad \text{5 places}$$

Positive exponent, so the answer is a large number.

**18.** $4.7 \times 10^{-8} = 0.000000047$

$$.00000004.7$$
$$\qquad \text{8 places}$$

Negative exponent, so the answer is a small number.

FIGURE 4

*Our conceptual presentation of a way to convert from decimal to scientific notation without complicated rules.*

*Bittinger, M.,* Introductory Algebra, *9th ed. Boston, Pearson/Addison-Wesley, 2003, pp. 308–309.*

encounters expressions like $13.95 \times 10^2$ and $0.8 \times 10^{-13}$, which are not in scientific notation. The reader usually has trouble knowing what to do. By simply converting the 13.95 to scientific notation $1.395 \times 10^1$ in the first case and then 0.8 to scientific notation $8.0 \times 10^{-1}$ in the second, and applying rules of exponents, the task becomes meaningful. It is shown in Figure 5.

## On Including the Answers

Do you want to open a firestorm of controversy? Then get one thousand instructors in a room and discuss whether the students should have all or half of the answers to the exercises at the back of the text. I assure you that if you cast a vote, the outcome of opinion will be 50% for and 50% against. Our solution was to put the answers to the odd-numbered exercises (about half)

**EXAMPLE 20**   Multiply: $(3.1 \times 10^5) \cdot (4.5 \times 10^{-3})$.

We have

$$(3.1 \times 10^5) \cdot (4.5 \times 10^{-3}) = (3.1 \times 4.5)(10^5 \cdot 10^{-3})$$

$$= 13.95 \times 10^2 \qquad \text{Not scientific notation. 13.95 is greater than 10.}$$

$$= (1.395 \times 10^1) \times 10^2 \qquad \text{Substituting } 1.395 \times 10^1 \text{ for 13.95}$$

$$= 1.395 \times (10^1 \times 10^2) \qquad \text{Associative law}$$

$$= 1.395 \times 10^3. \qquad \text{Adding exponents. The answer is now in scientific notation.}$$

**EXAMPLE 22**   Divide: $(6.4 \times 10^{-7}) \div (8.0 \times 10^6)$.

We have

$$(6.4 \times 10^{-7}) \div (8.0 \times 10^6) = \frac{6.4 \times 10^{-7}}{8.0 \times 10^6}$$

$$= \frac{6.4}{8.0} \times \frac{10^{-7}}{10^6}$$

$$= 0.8 \times 10^{-7-6}$$

$$= 0.8 \times 10^{-13} \qquad \text{Not scientific notation. 0.8 is less than 1.}$$

$$= (8.0 \times 10^{-1}) \times 10^{-13} \qquad \text{Substituting } 8.0 \times 10^{-1} \text{ for 0.8}$$

$$= 8.0 \times (10^{-1} \times 10^{-13}) \qquad \text{Associative law}$$

$$= 8.0 \times 10^{-14}. \qquad \text{Adding exponents}$$

FIGURE 5
*Sample annotated exercises converting numbers to scientific notation to multiply and divide.*

*Bittinger, M.,* Introductory Algebra, *9th ed. Boston, Pearson/Addison-Wesley, 2003, pp. 309, 311.*

and all the answers to the margin exercises, summary reviews, chapter tests, and cumulative reviews at the ends of the books.

## Writing Zeros before Decimal Points

We know that

$$0.2348 = .2348.$$

I remember driving from Lafayette, Indiana, to Minneapolis, Minnesota, with a group of mathematics educators when two of the profs, Grayson Wheatley and Robert Oesterle, debated the issue for what seemed like hours. This came to haunt us when Mike and I began writing. The question was, do we place the zeros before the decimal points or not? To us, the use of the zeros alleviated the possibility of forgetting or not reading the decimal point. When writing, if students included the zero, they were being sure the decimal point was included. Trite as this issue may seem, we got letters. The zeros are critical in medical notation. It'd be awful to give a patient 5 g of medication when it should be 0.5 g!

THE FAR SIDE® By GARY LARSON

**Hell's library**

© 1987 FarWorks, Inc. All Rights Reserved/Dist. by Creators Syndicate

The Far Side® by Gary Larson © 1987 FarWorks, Inc. All Rights Reserved. Used with permission.

## The Five-Step Problem-Solving Strategy

What is the most difficult mathematical task for students? What is the most difficult mathematical task for instructors to teach? And, what is the most difficult mathematical task for authors of developmental mathematics? You guessed it—word problems, word problems, word problems!! We preferred to call them "applied problems" because such wording implied that one could apply mathematics to a problem or application.

Problem solving started receiving great emphasis in mathematics education in the early 80s. Mike developed a five-step process for solving word problems (see Figure 6) during the second edition of our hardback algebra books and in the fifth edition of the trilogy. It is based in part on George Polya's strategy in his book *How to Solve It*.

As stated, the Familiarize step is probably the most important step. But, as well as this strategy has been received in our books, I know in my heart that this step is the one that students slight the most. To me, this is where students establish an experience base to a problem, using intuition. Yet, students simply do not want to spend time properly preparing themselves to do a problem. "Just tell me how to do it!" they say. My response is, "Spend more time getting experience with the problem and you will understand and be able to do it!"

Figure 7 is an example from our books, that gives extensive treatment of the Familiarize step before following through on the four other steps. It

# 2.6 APPLICATIONS AND PROBLEM SOLVING

### a   Five Steps for Solving Problems

We have discussed many new equation-solving tools in this chapter and used them for applications and problem solving. Here we consider a five-step strategy that can be very helpful in solving problems.

---

**FIVE STEPS FOR PROBLEM SOLVING IN ALGEBRA**

1. *Familiarize* yourself with the problem situation.
2. *Translate* the problem to an equation.
3. *Solve* the equation.
4. *Check* the answer in the original problem.
5. *State* the answer to the problem clearly.

---

Of the five steps, the most important is probably the first one: becoming familiar with the problem situation. The table below lists some hints for familiarization.

---

**TO FAMILIARIZE YOURSELF WITH A PROBLEM:**

- If a problem is given in words, read it carefully. Reread the problem, perhaps aloud. Try to verbalize the problem as if you were explaining it to someone else.

- Choose a variable (or variables) to represent the unknown and clearly state what the variable represents. Be descriptive! For example, let $L$ = the length, $d$ = the distance, and so on.

- Make a drawing and label it with known information, using specific units if given. Also, indicate unknown information.

- Find further information. Look up formulas or definitions with which you are not familiar. (Geometric formulas appear on the inside front cover of this text.) Consult a reference librarian or the Internet.

- Create a table that lists all the information you have available. Look for patterns that may help in the translation to an equation.

- Think of a possible answer and check the guess. Note the manner in which the guess is checked.

---

FIGURE 6

*A page from* Introductory Algebra *showing Mike Keedy's five-step process for solving word problems.*

Bittinger, M., Introductory Algebra, *9th ed. Boston, Pearson/Addison-Wesley, 2003, p. 185.*

comes from a dear friend, Dr. Edward M. Zeidman, formerly of Essex County Community College in Baltimore.

I recall seeing books that emerged during the early 80s with "problem solving" in their titles. Typically, at the beginnings of these books there was a greater emphasis on problem solving than had been in previous books. But after a few pages on the topic, the notion of using an overt strategy was dropped and the problem solving reverted to the methods of the past. Mike and I committed ourselves to the notion that we wanted to include a strategy, establish it to our satisfaction, and continue it throughout all our books.

---

**PROBLEMS INVOLVING WORK**

**EXAMPLE 1** *Recyclable Work.* Erin and Nick work as volunteers at a community recycling depot. Erin can sort a morning's accumulation of recyclables in 4 hr, while Nick requires 6 hr to do the same job. How long would it take them, working together, to sort the recyclables?

**1. Familiarize.** We familiarize ourselves with the problem by considering two *incorrect* ways of translating the problem to mathematical language.

a) A common *incorrect* way to translate the problem is to add the two times: 4 hr + 6 hr = 10 hr. Let's think about this. Erin can do the job alone in 4 hr. If Erin and Nick work together, whatever time it takes them should be *less* than 4 hr. Thus we reject 10 hr as a solution, but we do have a partial check on any answer we get. The answer should be less than 4 hr.

b) Another *incorrect* way to translate the problem is as follows. Suppose the two people split up the sorting job in such a way that Erin does half the sorting and Nick does the other half. Then

$$\text{Erin sorts } \frac{1}{2} \text{ the recyclables in } \frac{1}{2}(4 \text{ hr}), \text{ or 2 hr,}$$

$$\text{and} \quad \text{Nick sorts } \frac{1}{2} \text{ the recyclables in } \frac{1}{2}(6 \text{ hr}), \text{ or 3 hr.}$$

But time is wasted since Erin would finish 1 hr earlier than Nick. In effect, they have not worked together to get the job done as fast as possible. If Erin helps Nick after completing her half, the entire job could be done in a time somewhere between 2 hr and 3 hr.

We proceed to a translation by considering how much of the job is finished in 1 hr, 2 hr, 3 hr, and so on. It takes Erin 4 hr to do the sorting job alone. Then, in 1 hr, she can do $\frac{1}{4}$ of the job. It takes Nick 6 hr to do the job alone. Then, in 1 hr, he can do $\frac{1}{6}$ of the job. Working together, they can do

$$\frac{1}{4} + \frac{1}{6}, \text{ or } \frac{5}{12} \text{ of the job in 1 hr.}$$

In 2 hr, Erin can do $2\left(\frac{1}{4}\right)$ of the job and Nick can do $2\left(\frac{1}{6}\right)$ of the job. Working together, they can do

$$2\left(\frac{1}{4}\right) + 2\left(\frac{1}{6}\right), \text{ or } \frac{5}{6} \text{ of the job in 2 hr.}$$

Continuing this reasoning, we can create a table like the following one.

| TIME | FRACTION OF THE JOB COMPLETED | | |
|---|---|---|---|
| | Erin | Nick | Together |
| 1 hr | $\frac{1}{4}$ | $\frac{1}{6}$ | $\frac{1}{4} + \frac{1}{6}$, or $\frac{5}{12}$ |
| 2 hr | $2\left(\frac{1}{4}\right)$ | $2\left(\frac{1}{6}\right)$ | $2\left(\frac{1}{4}\right) + 2\left(\frac{1}{6}\right)$, or $\frac{5}{6}$ |
| 3 hr | $3\left(\frac{1}{4}\right)$ | $3\left(\frac{1}{6}\right)$ | $3\left(\frac{1}{4}\right) + 3\left(\frac{1}{6}\right)$, or $1\frac{1}{4}$ |
| $t$ hr | $t\left(\frac{1}{4}\right)$ | $t\left(\frac{1}{6}\right)$ | $t\left(\frac{1}{4}\right) + t\left(\frac{1}{6}\right)$ |

From the table, we see that if they work 3 hr, the fraction of the job completed is $1\frac{1}{4}$, which is more of the job than needs to be done. We see again that the answer is somewhere between 2 hr and 3 hr. What we want is a number $t$ such that the fraction of the job that gets completed is 1; that is, the job is just completed.

**2. Translate.** From the table, we see that the time we want is some number $t$ for which

$$t\left(\frac{1}{4}\right) + t\left(\frac{1}{6}\right) = 1, \quad \text{or} \quad \frac{t}{4} + \frac{t}{6} = 1,$$

where 1 represents the idea that the entire job is completed in time $t$.

**3. Solve.** We solve the equation:

$$12\left(\frac{t}{4} + \frac{t}{6}\right) = 12 \cdot 1 \qquad \text{Multiplying by the LCM, which is } 2 \cdot 2 \cdot 3, \text{ or } 12$$

$$12 \cdot \frac{t}{4} + 12 \cdot \frac{t}{6} = 12$$

$$3t + 2t = 12$$

$$5t = 12$$

$$t = \frac{12}{5}, \text{ or } 2\frac{2}{5} \text{ hr.}$$

**4. Check.** The check can be done by recalculating:

$$\frac{12}{5}\left(\frac{1}{4}\right) + \frac{12}{5}\left(\frac{1}{6}\right) = \frac{3}{5} + \frac{2}{5} = \frac{5}{5} = 1.$$

We also have another check in what we learned from the *Familiarize* step. The answer, $2\frac{2}{5}$ hr, is between 2 hr and 3 hr (see the table), and it is less than 4 hr, the time it takes Erin working alone.

**5. State.** It takes $2\frac{2}{5}$ hr for them to do the sorting, working together.

FIGURE 7

*An example of the five steps to problem solving.*

*Bittinger, M.,* Introductory Algebra, *9th ed. Boston, Pearson/Addison-Wesley, 2003, p. 511–512.*

### Division of Labor

Mike Keedy and I had a great run together as co-authors. Our work together lasted from 1968 to 1984. As we became more and more successful, Addison-Wesley signed us to do many books for both high school and college. In 1984 Mike and I determined that our stable of textbooks was large enough that it warranted separating our writing duties, he taking over the high school books and I the college books. This made the volume of our workload more reasonable and, indeed, my passion was much more for college students. I will always be indebted to Mike Keedy for the lasting impact he has had on my career.

## NEW CO-AUTHORS

In the years following my work with Mike Keedy the books moved to four colors and there was exponential growth in the number of supplements for our books. But most notably, those years involved growing relationships with four tremendous co-authors: Judy Beecher, David Ellenbogen, Judy Penna, and Barbara Johnson.

### Judith A. Beecher

When Mike Keedy and I stopped working together it was clear to me that I could not handle the stable of books that needed my attention. Judy Beecher, my first real "find," had worked for Mike and me checking manuscripts, writing student solutions manuals, and writing printed test banks. Her background included high school teaching and part-time teaching at Indiana University–Purdue University at Indianapolis (IUPUI). It was at IUPUI that I met Judy when she came to me for advice on her adjunct teaching. Since I was always on the prowl for qualified people to accuracy-check manuscripts, I started her out in that process and soon moved her to other tasks.

She became so close to my work that she soon became what is called in the publishing business a *developmental editor,* a special individual who reads reviews and various drafts of the manuscripts and acts as an advisor, bouncing ideas back and forth with the authors. When I am struggling about what to do with a topic, the first person I consult is Judy; the second is usually my editor at AW. Judy contributes a unique brand of creativity as she reads and reviews a manuscript. I think of her as my "write-arm."

I usually do a "first draft," and then it goes to Judy to read and mark up. Then it comes back and I do a "Judy Draft." After that I submit the manuscript to the publisher for review.

Judy has been so dedicated and loyal to our work that many times she took manuscripts to her son's baseball games. On other occasions, when traveling to Ohio to visit family, she would sit in the back of the car and do book work while her husband Art drove. Judy has always been a sounding board, a trusted confidant, a loyal co-author, and most of all a friend. I treasure our working and professional relationship. What would I do without my Judy? It is a question I never want to answer.

*The artist is nothing without the gift, but the gift is nothing without the work.*
—EMILE ZOLA, 1840–1902

If you read my list of publications, you will notice that I diverged from my math books once to write a book (with Dusty Baker and Jeff Mercer) entitled *You Can Teach Hitting.* Since Judy knows little or nothing about playing baseball, it never occurred to me to involve her with that book. But, as fate would have it, she could not stand to see something with my name on it not cross her desk, and she asked to help.

Luckily, I consented, and between Judy and my sister-in-law, Valerie Bittinger, the book benefited from having two women read the book who knew nothing about coaching the game. The goal was to create a book that any single parent could use to teach a child or an adult to hit. With Judy's help, we met the goal.

### David J. Ellenbogen

I first met David after my editor at the time, Stuart Johnson, had discussed the notion of finding someone to help me co-author the hardcover series *Elementary Algebra: Concepts and Applications* and *Intermediate Algebra: Concepts and Applications.* These hardcover versions were adaptations of our paperbacks, but because the workspace and margin exercises were removed, we could add a bit more rigor and lots more exercises. David was brought in during the writing of the third editions.

Stuart and I had decided to audition two instructors as co-authors. We asked them to rewrite the same chapter of one of the texts. When the auditioning manuscripts came back, we quickly rejected the one enclosed with a note saying in effect, "I could have done this project much better, but I want to write my own book some day, so I did not do all I could have." We thought he knew that we were offering him his own book. David, on the other hand, poured his whole heart and soul into that manuscript, and the quality showed.

David served as a consultant to me at first as I revised the elementary algebra book. We would both read the reviews of the second edition, then we would talk about each chapter or topic before I revised it. He would then read my first draft. When it was time to revise the intermediate algebra book, David became the primary author and I the consultant. After sending me the first draft for review, we had many telephone conversations, so many that I sometimes wondered if I would ever have time to write anything else. But the process was well worth it because the third editions were accepted better than ever, and those books became a driving force in the hardback algebra market.

As time went on, David became the force behind writing *Prealgebra,* a book we adapted from *Basic Mathematics.* It contains negative numbers at the beginning with some algebra integrated throughout the book. This book was a bit higher in level than basic mathematics and is intended to better prepare students for the hardback algebra books. David was also a co-author of the first edition of our hardback graphing calculator algebra-trig series.

David is a dedicated instructor at the Community College of Vermont and that dedication extends to his writing. He is never reluctant to do extra rewrites or to put in the hard work to improve a book. I have always admired his passion to find applications. He is very good at creating interesting applied problems covering a range of items as diverse as floor stain mixing, octane ratings, zebra mussels, global warming, and train schedules in France.

### Judith A. Penna

Judy was a part-time instructor at IUPUI when I met her. She started her work with us in much the same way as Judy Beecher had. As Judy Beecher's role changed from doing solutions manuals and textbooks to being a developmental editor and author, Judy Penna took up that task, eventually becoming the coordinator of all the Bittinger supplements.

When I think of Judy P, I think of "trust." With her at the helm of the supplements, I never worry about their quality and reliability. When she supervises a videotape shoot, I know it will run well and the tapes will be of high quality. Her supervision of supplements has allowed me to write in other areas and to begin cutting back on my workload at this point in my life.

It was inevitable that Judy P would consider becoming a co-author. At first she preferred the security of focusing on supplements, but finally asked for an opportunity to be a co-author. That began with our graphing calculator algebra-trig series and eventually led to a grapher-optional algebra-trig series.

### Barbara L. Johnson

Barbara's route to our writing team was a bit different. As a teacher at Colonial Christian School in Indianapolis, Barb was my son Lowell's algebra-trig instructor. Lowell was quite complimentary about her teaching. I soon met her at basketball games and decided to ask if she'd be interested in accuracy-checking manuscripts. When she excelled at that task, we gave her more work, eventually asking her to write some fresh application problems. When these turned out to be of high quality and very creative, we invited her to do some writing with David Ellenbogen on his hardback books. Those books have been well received.

Being a successful author requires heart, soul, and passion. The task is too demanding for a half-hearted effort. Writing is hard work. I am proud of our successful author team. We work together well in an atmosphere of trust and forthrightness. We are never afraid to tell one another of some difficulty in our writing. We love our work and are truly dedicated to the students.

## CREATIVITY AND THE DRIVE TO WRITE

Psychologists assert that humans are creative in ways that no other animal species or even a computer can be. As an oversimplified illustration, look at the backsides of the pieces of a puzzle in Figure 8. A computer can be programmed, with the help of a robotic arm, to eventually put the backsides of the puzzle together, as in Figure 8B, but no computer can go the next step,

FIGURE 8

in Figure 8C, and bring the next dimension, the beautiful picture of the scenery.

The next dimension is creativity, going beyond simply putting the pieces of a puzzle together and actually bringing a fresh outlook or explanation to a situation. In the case of a mathematics textbook, the creativity might show up in discovering a better way to explain a topic to students, a new strand of exercises, or a new design for layout of the pages.

Can you teach creativity? I answer this question by comparing my bowling and golf games. I took the time and had the passion to improve my bowling game, but not excel as professional bowler. My golf game is another issue. I have played the game for over twenty years and have taken some instruction, but I have never had the passion to involve myself in the practice and lessons it would take to move my game to the next level. I was a 115 score golfer some twenty years ago and, although I now score around 105, I don't consider the change much of an improvement. I had the talent and passion to improve my bowling, but I do not possess the talent or the will to improve at golf.

I did have an overwhelming will and passion to be an author. Anyone can be taught to be better at a task, but is the ability to learn or improve skills a special gift? While many people can be taught to be a more creative textbook author, I believe to excel is a "spiritual gift."

M. Scott Peck, author of numerous best-selling titles, most notable of which are his three *Road Less Traveled* books, distinguishes between two types of mentality: the "secular consciousness," and the "sacred (spiritual) consciousness" when considering gifts and creativity.

*An individual with a* secular *consciousness considers himself or herself to be the center of the universe. Such individuals are often very sophisticated. They realize that every other individual considers himself or herself to be the center of the universe. They also realize that they are but single individuals among five [now six] billion of us human beings scratching out an existence on the*

REFERENCE
Peck, M. S., and von Walder, M., *Gifts for the Journey*. Renaissance Books, 5858 Wilshire Blvd., Suite 200, Los Angeles, CA, 90036, 1999, pp. 119–120.

*surface of a small planet, circling a medium-sized star among millions of stars in but one of countless galaxies. So even though an individual with secular consciousness considers himself to be the center of things, he or she is often afflicted with a sense of meaninglessness.*

*The situation is exactly the opposite for those with a* sacred *consciousness. Here the individual does not consider himself or herself to be the center of the universe at all; the center of the universe is God. Rather than suffer from a sense of meaninglessness that results from the belief in the centrality of the ego, however, individuals blessed with the sacred consciousness find their lives filled with meaning by nature of their relationship to the true center. God gives them meaning. So, egoless in one way, they paradoxically know themselves to be of great importance in the scheme of things—an importance derived from God.*

*The sacred consciousness is wisdom. It is salvation. The quest for wisdom is the quest for transformation—the conversion of secular consciousness into, and even deeper into, the sacred consciousness. (pp. 119–120)*

In another book, Peck discusses consciousness and originality:

*What is infinitely more mysterious is the origin of originality. Where does a fresh vision come from? A startlingly different imagination? What creates a creative genius? . . . J. S. Bach explained, "God writes my music." The same can be said of all great art. Artists are no more religious than other people. When they are secular-minded, they have a notable lack of desire to discuss where their art came from. They suddenly turn prosaic, saying, "So and so taught me this technique" and "So and so taught me that one," as if the matter of inspiration, an influx of the Spirit was irrelevant. When they are religious-minded, they will simply agree with Bach, but they will be quiet about it, because talking about such things in society can get one in trouble.*

*. . . about all we can do at the end of the matter is shrug our shoulders and change the subject by lamely proposing, "I guess it is a gift."*

I believe that being a good textbook author is a spiritual gift. C. S. Lewis said:

*But to speak of the craft [of writing] itself, I would not know how to advise a man how to write. It is a matter of talent and interest. I believe he must be strongly moved if he is to become a writer. Writing is like a "lust," or like "scratching when you itch." Writing comes as a result of a very strong impulse, and when it does come, I for one must get it out.*

It is not a matter of money, or power, or success; it is a matter of the itch, the lust, the passion. My writing is driven by my passionate desire that students learn.

As Peck says, "In each case, I've been able to keep my nose to the grindstone only because I felt 'called' to write that particular work. I'm not suggesting that I'm a great artist or a godly man. I'm only suggesting that I didn't have that much choice about it. And that many authors or creators have

REFERENCE
Peck, M. S., *In Search of Stones.* New York, Hyperion, 1995, pp. 356–358.

REFERENCE
Lewis, C. S., "Cross Examination," *God in the Dock.* Grand Rapids, Eerdmans, 1970, pp. 258–267.

REFERENCE
Peck, M. S., *In Search of Stones.* New York, Hyperion, 1995, pp. 355–358.

*Don't aim at success—the more you aim at it and make it a target, the more you are going to miss it. For success, like happiness, cannot be pursued: it must ensue . . . as the unintended side effect of one's personal dedication to a course greater than oneself.*

—FRANKL, V., *Man's Search for Meaning*, New York, New York, A Washington Square Press Publication of Pocket Books, 1984, pp. 16–17

*Everyone who does not* need *to be a writer, who thinks he can do something else, ought to do something else.*

—GEORGES SIMENON

Luke 12: 15 *". . . Watch out! Be on guard for all kinds of greed. A man's life does not consist in the abundance of his possessions."*

—All Biblical quotes are taken from the New Living Translation of the Bible, Thomas Nelson Publishers, Nashville, Tennessee, 1998

REFERENCE

Morris, T., *If Aristotle Ran General Motors: The New Soul of Business.* New York, Henry Holt, 1997. p. 94.

found themselves in the same sort of unpleasant predicament of operating 'under orders.'"

I will never forget the story of a potential author that my editors told when I started writing. He shall remain nameless, but he signed with a publisher to do a vast series of math textbooks ranging from liberal arts math to algebra-trig through calculus. He received a huge advance against royalties. I do not know the amount, but I do know that his wife spent it all on a Mercedes-Benz. What was the result? His first two books failed, and the series was never completed. I question his motivation—was it to make money, or was it to help students learn? Was is it a secular or a sacred consciousness?

Suffice it to say that, whatever your vocation, if money, power, and success are your primary motivators, you are responding to the wrong motivation. A question you might ask is, "Would I do this for free?" If your answer is "Yes," you are well on the road to being a success.

I have often been enticed by publishers to write a "big" calculus book—the kind used by math, science, and engineering majors that sells hundreds of thousands of copies. Yet a still, small, voice says, "You do not have the passion or the classroom experience to tackle such a project! You would only be doing it for money, power, and success. Plus, you wouldn't be creating; you would be competing. Don't do it!" Indeed, such books are over 1,000 pages and take four or five years to write. The strain on my family and me would have been damaging at best. It might be hard to believe I ever say "No," given so many publications, but I do turn projects down, especially the "big" calculus text.

## Have You Loved Enough?

Tom Morris is the author of self-help books for business people. In one book he tells the story of a long seminar he gave, during which the attendees kept pressing him about the meaning of life. He eventually gave them the following definition, "The meaning of life is creative love. Loving creatively."

I think that definition also applies to a successful author. Is the writing an act of creative loving? When I sit down at the computer to create math books, I imagine that I am next to the student thinking, "How can I present this to you so you can learn? Let's try this." When I meet with my co-authors and editors and we grapple with writing issues, I always ask, "Can these students understand if we present the math in this manner?" I believe the decision is an act of loving creatively.

Peck supports this as well saying,

Love *is the second most compelling ingredient in the creative process, hardly separable from the first. [The first being his spiritual consciousness] The subjective judgment "it is good" needs to be tested by the objective effect of the art*

REFERENCE

Peck, M.S., *In Search of Stones.*
New York, Hyperion, 1995,
pp. 356–358.

*upon at least a few other human beings. Granted that taste is ephemeral, if you create a work that is not deeply appealing to a single other person in your lifetime, it is probably not because your opus is timeless; likely it is because you have not loved enough, and your sole assessment that your work is "good" is something that is only and forever, like Narcissus' fatal attraction. . . . when I write I do my best, within constraints, to bear my reader in mind, and the successful artists I know do likewise. Frankly, it is a bit erotic. I am ever mindful of my beloved [the reader or student]."*

In summary, the creative process is a result of the union of one's spiritual gifts and the passion one applies to use them.

$$\text{Creativity} \; = \; (\text{Spiritual Gifts}) \bigcup (\text{Passion}).$$

## THE CRAFT OF WRITING

A typical day in my life as an author might go as follows.

- 6:30 A.M., rise, breakfast, papers, read,
- 8–11 A.M., work write,
- 11:45 A.M.–12:15 P.M., rest,
- 12:15 P.M.–3:30 P.M., exercise, run errands,
- 3:30 P.M.–5:00 P.M., work, write,
- 5:00 P.M.–5:30 P.M., dinner,
- 5:30 P.M.–7:00 P.M., work, write.

Since my heart attack in 1998, I have had to eliminate many parts of my diet, but I have been able to maintain my coffee, Green Mountain's Rainforest Nut is my favorite. Following breakfast, I browse parts of the local paper, and *USA Today,* always on the prowl for interesting applications and articles on education. Most of the time I hone in on the sports and entertainment sections. Then I check my e-mail and copy articles off the Internet to read later while I exercise.

It is important for me to spend time reading something other than the newspapers or mathematics. I read the Bible and other Christian theology, either in the morning or while I am on the treadmill in the afternoon, or in the middle of the night when I can't sleep. I am amazed how often I get ideas about my math books during this time; it is like they come between the lines of my reading. Some might call this the listening part of prayer. Philosophical writers M. Scott Peck and Philip Yancey suggest that 80 percent of their prayer time is spent in listening. During these times it seems God feeds me ideas about my work, which almost always lead me to anticipate my writing. Perhaps it is really God urging me on. So as not to make this too ethereal, I confess that there are certainly times when I am flat or dull and don't want to write . . . I am in the valley of despair, or having writer's block.

Today, I might be writing Chapter 8, *Systems of Equations*, in the ninth edition of *Introductory Algebra*. In contrast with writing about radical expressions, I like this topic because all the notation is linear and easy to keyboard. Yesterday, I looked at the ten reviews of this chapter and transferred the suggestions that I considered worthy of possible change to a working copy of the book. Sometimes reviewers send marked-up books as well. As I go through the marked-up chapter, page by page, I reconsider the changes and prepare the manuscript. When they seem reasonable and short, and easy to insert, I usually do them, especially when they do not contradict my instincts as an educator and when page count is not affected. When a suggestion involves more pages or may be more global, affecting the entire book, then I grapple with the issue. Sometimes I call my co-author, Judy Beecher, and we discuss it further. Usually, Judy and I reach a conclusion using the following criteria.

- How many pages will it take, and do we have the page count to allow the changes?

- Is this a reviewer with an isolated stand on a particular issue? If the comment comes from several reviewers, we give it stronger consideration.

- Have our editors pushed for the change?

- Will the students understand the change, and will it be a positive instructional device?

In the final analysis, an author must trust his instincts as an instructor and writer to make a decision.

I usually work from say 8 A.M. until lunch time at 11 A.M. But, almost always I eat at my desk and keep going until about 11:45 A.M. Then I lie down for a short nap or rest in front of the TV. At 12:15 P.M., I leave to exercise. I go to a health club and walk–run about 5 miles on the treadmill. This takes roughly another $2\frac{1}{2}$ hours out of my day, but I deem it a necessity in my life since I had a heart attack in 1998. I try to offset the effects of coronary artery disease by a careful diet and extensive exercise. I return to my desk about 3:30 P.M. and work until supper. After supper I work until about 7 P.M., and become a couch potato until about 10 P.M., watching TV with Elaine. When I was younger, I might work until 10 P.M.

I am somewhat of an insomniac. If I wake up in the night and can't go back to sleep, I sometimes get up and work for an hour or two. I certainly don't have distractions. It can very well be the case that I can't sleep because I mentally take my work to bed and have some topic on my mind. Finally, it drives me to get up and put it down on paper.

### The Writing Plan

Usually we establish a plan for a revision. It results in a checklist of tasks to carry out with each chapter, tasks such as write more study skill inserts, or more technology exercises, or more skill maintenance exercises. I go through the checklist and complete the tasks for each chapter.

Applied (word) problems usually are an adventure unto themselves. It is well known that students dislike them because of the time and effort involved. It is perhaps a well-kept secret that most instructors dislike having to teach them and, as an author, I dislike having to write them for the very same reasons, time and effort. Applied problems take a long time to research and to write. But, it is indeed a satisfying part of writing when I can come up with a worthy method of writing about problem solving. If only students would take the time to apply the method!

When the main sections of a chapter are finished, I turn to writing the review exercises and chapter test. The chapter is sent off for more reviews which lead to more rewrites. When the next set of reviews has been considered, another draft is prepared. As far as the exercises, the ideal situation is for the author to prepare his own answers. Lots of worthwhile polishing can then occur. The author knows the intent of a problem. For example, maybe an applied problem has a fractional answer, when only a natural number makes sense. The list of possibilities for mistakes is endless—I have made most of them.

Then the entire chapter is checked for accuracy by two or three people. When their work is complete, we check their corrections and at last we have a final draft. We never incorporate a change without double-checking.

Time is always a big factor in a revision plan. I know of one former author, not particularly successful, who would grind out $n$ pages each day, no matter the quality. I do not recommend this trap—three quality pages a day is better than twenty of poor quality. I have been known to get a chapter done in an hour, but it was probably part of a revision that required very little work. In other situations, I may spend an entire day or even two on one applied problem. It is typical to spend a week or two per chapter, and eight to twelve months per book.

If you are considering being an author, and trying to get started, pick a topic that seems easiest to you or one you are very excited about. This can minimize early road blocks that might discourage you from completing your book.

### Interruptions

It may surprise my wife and family to read this, and I think that I have excelled at covering it up, but I hate to be interrupted when I am writing!

The notion of *flow* as described by Mihaly Csikszentmihalyi, in his book *Flow: The Psychology of Optimal Experience* explains the problem:

> *". . . flow—the state in which people are so involved in an activity that nothing else seems to matter; the experience is so enjoyable that people will do it even at great cost, for the sheer sake of doing it."* Sports addicts would call flow being "in the zone."

REFERENCE
Csikszentmihalyi, M., *Flow: The Psychology of Optimal Experience*, New York, HarperCollins,1990, p. 4.

Any interruption, whether it's as blatant as the phone ringing or as subtle as Elaine walking into the room to use the computer or file a document, can break my flow. It takes effort to get it back, effort I would rather not expend. A worthy idea can sometimes be lost in the process.

*Donald Trump, a busy and successful businessman if ever there was one, always allows himself to be interrupted when a phone call comes from one of his children.*

*One of my granddaughters, Claire E. Bittinger, a beloved interruption on top of my desk.*

REFERENCE

Peck, S., *Golf and the Spirit*. New York, Three Rivers Press, New York, p. 5.

Baker, D., Mercer, J., and Bittinger, M., *You Can Teach Hitting*, Indianapolis, Bittinger Books, 1993, www.dustybaker.com.

*Creating is not indulging the self; it is denying the self. Self will be seen to be the greatest block to true artistic expression.*

—MICHAEL CARD, author and performer of inspirational music

Nevertheless, it has always been my objective never to become upset when disturbed by Elaine, or my children, especially when they were younger. I will never forget the little fingers of my sons on the edge of my desk and the eyes looking up at me longingly with the words, "Dad, do you want to play catch?" I always stopped—I never said "No." They still recall the experience, so I must have done something right.

I love working at home, but it can be both a blessing and a curse; a blessing because it is so easy to get to work, especially when the weather is bad and a long commute can be avoided, and a curse when the interruptions occur.

The extreme interruption occurs when an entire day is lost. They don't occur very often, but there have been enough lost days over the years to get my attention. On such days, one phone call leads to another, some get stacked up, and I do no creative writing. Another lost day involves encounters with the "piddlys" on my desk. When I travel and return home, I not only have to deal with my expense receipts but also a mountain of mail of all types: bills, letters that need attention, and phone calls that must be returned. Inevitably there is a crisis to be dealt with.

### Kenosis

M. Scott Peck defines the concept of *kenosis* as the "process of emptying oneself of self." (p. 5) In bowling, golf, baseball, or probably any sport, kenosis comes to bear in the simplistic command, "Relax!" By emptying oneself of self and just letting your arm swing, your bowling will improve. Similarly, Dusty Baker, in our book, *You Can Teach Hitting*, asserts that the best kind of baseball swing occurs when the batter holds the bat in his fingertips and throws it at the ball in a tension-free, relaxed, manner. And perhaps Sarah Hughes, the 16-year-old champion figure skater of the 2002 Olympics, experienced kenosis when she performed one of the most startling upsets in Olympics history, moving from fourth place to the championship in the final long program. She threw caution to the wind, trusting her talent and love of skating. Sarah said, "I skated for pure enjoyment."

You might be thinking, "A math textbook is filled with lots of thought processes and sequences of steps. How can you turn off the self in writing such a book?" The answer concerns not so much the mental demands of doing math, as the conceptual process of creative writing. Following are three examples.

I have had the "Aha!" experience so often that it has become a common reality in my life. In solving a difficult problem, I often have to grapple with it for a long time. As a textbook author, my problem is usually trying to find the best way to present a concept. Now at the age of 62, I experience the usual male problem of having to rise from my sleep to go to the bathroom. Amazingly, I often get up around 3:30 A.M. Inevitably, my mind starts back to work, and "Aha!" I think of a creative solution to my problem. Sometimes I go downstairs and type it out on the computer.

I have analyzed this experience many times and discussed it with friends and family, but I have finally decided that it's a God-thing. Recently, I was blessed to have a long conversation with M. Scott Peck, who informed me that these 3:30 A.M. "Aha!" experiences may be a God-thing, but it is also kenosis. When one sleeps, the mind turns off the e-mails, the TV, the routine of the day, and the needs of the family. In short, the mind is emptied and able to hone in on the problem, and it gets solved. This understanding of kenosis shed so much light on the workings of creativity for me that it was an **MPX**.

**MPX** MUSIC: "Dreamers," by Sarah Brightman, Composer Marvin Hamlisch, Lyrics by Christopher Adler, from the musical *Jean Seberg*, Track 5 on the CD *The Songs That Got Away*, recorded by Sarah Brightman, The Really Useful Record Co., Ltd., London.

MOVIE: *My Heroes Have Always Been Cowboys*, (1991) starring Scott Glenn and Kate Capshaw. The scene where Glenn talks about riding the bull and what it means to him expresses for me the challenge of a long hike, a baseball game, a bowling tournament, or writing a math book.

After learning more about kenosis, it dawned on me that this occurs in other situations. I analyze a topic, I try to write about it creatively, but it just does not seem right. I even think if I sit over it long enough, the answer will come. Eventually, I decide to just get away from it. I read, workout in the afternoon, bowl, go to a sporting event, or just watch TV. While my conscious attention is on these other activities, the subconscious remains fixed on the problem. I free myself of the conscious pressure to solve the problem, but the subconscious keeps working. When I do go back to it, the answer often jumps right out at me.

### Peaks and Valleys

I don't always have a burning desire to get to my computer and write. Indeed, I am often drawn away by the TV, the newspapers, and the Internet. On Sunday mornings we usually leave for church at 8 A.M. Before then, I read the paper, check my e-mails, and read sports articles on the Internet. After my sixth cup of coffee, I often feel drawn to the computer to occupy my time until we leave. Inevitably, I get into a flow on my work and when it is time to leave, I don't want to stop. This happens almost every Sunday and I have wondered, "Has God blessed my work because He knows I am about to go to church?" Perhaps. From Peck I learned that this is another kenosis situation. By thinking of the interruption of church, I have partially emptied myself of life's daily distractions, and I enter a flow of work.

*Writing is easy; all you do is sit staring at a blank sheet of paper until the drops of blood form on your forehead.*

—GENE FOWLER

Writer's blocks, or valleys of despair, can also occur. I normally pass them off as time spent "writing books in my mind." Sometimes, no matter how much thought I give to a topic, nothing seems to work. Many times, I just have to make myself sit down and "do something." Once started, I get into a flow, and by the grace of God, something worthwhile usually comes out. I think about my writing when I'm at the movies, watching TV, driving, and, embarrassingly, when listening to my friends and loved ones. No matter how hard I try, I seem to have such a passion for my work that I can't turn it off. I want to thank my friends and loved ones for forgiving my poor listening skills.

I often carry note cards or a small voice recorder to record ideas when I'm away from my desk. Ideas are like gold nuggets—hard to come by but easy to lose. There have been times when I was sure I would remember an idea but have lived to regret losing it. Now I pull off to the side of the road to write down an idea.

Clearly, there are peaks and valleys in the life of an author. I have been in flow and really enjoying writing on a topic, feeling the exhilaration of getting it done—a peak. Then I go to a new topic and slump into a "valley of despair," or "writer's block."

One would think that the experience of writing for thirty years would lead me past these points of desperation. But with new material, I sometimes get to feeling I will just never figure out what to do—I get the blues. Inevitably, after talking to my co-authors or editors, making some false starts, compiling little note sheets of ideas, and trying other techniques I've described in the past two sections, I start to climb out of the mire and get the job done.

Then I can sink right back into another valley. I can't help but think of my creative endeavors riding along a trigonometric curve like the following.

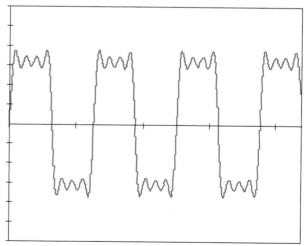

## VIEWS OF OTHER AUTHORS

To explore other viewpoints about the creative aspects of writing, I have either interviewed or researched tips on creativity from other authors. I wish

*I give the impression of knowing far more than I do because I work hard on research, write glibly, and keep extensive files of clippings on everything that interests me.*

—MARTIN GARDNER, quoted by Kendrick Frazier in an interview published in the *Skeptical Inquirer*, March, 1998. Gardner is known for his writings on mathematical puzzles and books on science.

REFERENCE

Sayers, D., *The Mind of the Maker.* Harper San Francisco, reprint edition, 1987.

Yancey, P., *Reaching for the Invisible God.* Grand Rapids, MI, Zondervan, 2000, pp. 124–125.

to thank them all for their contribution to this section. Although only one is a textbook author, the tips are valid for all authors. You will see a certain similarity in these comments. As an example, I once read the following stages of writing a book, though I do not recall the source:

1. Worrying
2. Planning
3. Writing
4. Revising (more than once)
5. Proofreading

I would add that stage 1 might be thought of as loading up the subconscious and/or doing research and that stage 4 requires the most work and perseverance.

### Philip Yancey, Author of Theological Nonfiction

Phil has a number of excellent, thought-provoking books on Christian theology. I like Yancey because he is so honest and transparent about the spiritual struggles in his life. He also has a way of helping to find a fresh path out of my spiritual doubts and dilemmas.

In discussing the stages of his writing, Yancey makes reference to the work of Dorothy Sayers, asserting the following three stages of creative art:

1. Idea
2. Expression for the idea
3. Recognition

Yancey elaborates as follows in terms of his book *Reaching for the Invisible God*:

*1. An Idea: Every writer begins with an Idea. Consider this book for instance. For several years, I read other books, talked to people, and scribbled notes on scraps of paper relating to a vague Idea. I had no title in mind, no clear concept of shape the book might take, only a strong desire to explore my own questions about how we visible humans can relate to an invisible God. Sometimes friends would ask, "What are you working on, Philip?" and I would try to explain, but their blank looks told me that my original idea was impenetrable.*

· · ·

*Most writers I know have a mild panic attack when asked the question, "What are you working on?" I [Yancey] want to reply, "I don't know yet. Let me finish writing it, and then I can tell you." The idea exists only as a first stage in the creative process.*

*2. Expression for the Idea: My [Yancey's] expression changes shape daily. Just yesterday I moved a huge block of text from one chapter to another, then deleted several pages entirely. On average I end up cutting a hundred pages from the first draft of each of my books. As I edit, I realize that some pages over which I labored many days disrupt the original Idea by causing the book to bog down or go in conflicting directions. The Idea has a life of its own, and over time I have learned to follow instincts alerting me when my Expression misrepresents the Idea. Similarly, my friends who write fiction tell me that the story itself leads them in ways they neither planned nor anticipated. Regardless of the medium, every human creator seeks to express the Idea perfectly, and falls short. When Michelangelo visited the Sistine Chapel after its completion, I'm sure he noticed every flaw and imperfection.*

*3. Recognition: The act of creation does not end, though, when I finish the work: another person must receive it. An artist creates for one purpose, to communicate, and the creative process will remain unfinished until at least one other person receives it.*

*A successful work of art summons up a response in the receiver. In fact, when we encounter great art something akin to a chemical bonding takes place and our very bodies respond: muscles, heart rate, breathing, perspiration. Playwright Arthur Miller said he never relaxed until he sat in the audience and looked in people's eyes. If he saw the spark of recognition—"My God, that's me!"—he knew he had succeeded with his play. Recognition completes the cycle of creativity.*

I must say on a less ethereal note that I agree with Yancey's words. The *idea* for a textbook might be some fresh way to present the material, the *expression* is carrying out the idea in a consistent manner throughout the text. And the *recognition* by the instructor and/or student can be extremely satisfying and is a critical stage of the creative process. I love to know that a student's muscles, heart rate, breathing, and perspiration might respond to one of my math books; I am not sure that has ever happened, although there have been many kind letters and e-mails over the years. One would not keep writing if this Recognition stage was not satisfied.

And, yes, some ideas fail to pass the Recognition stage. You are probably not aware of my unsuccessful books, which I will just allow to fade into oblivion. I will tell you about one *idea* that failed in the ninth edition of the Trilogy. I had this wonderful idea to categorize the learning objectives as Manipulation, then Solving Equations, and then Problem Solving; a sort of stair-step hierarchy that evolves in almost every chapter of our algebra books. This idea failed because not only could I not get it across to the editors, but the reviewers failed to understand it clearly. If the editors don't understand it, they can't explain it to their sales reps, and the reps can't explain it to potential adopters. I ditched the idea, but I still think it has value.

Having read several of Yancey's books, I was fortunate to be able to sit down with him and discuss his creative efforts in more detail. Yancey quotes from numerous sources in his books, so much so that I asked him if he took his note cards on a particular subject and pored over them until he came up with a writing plan that was new and fresh. He confirmed my suspicions, but added that he intentionally creates a very detailed writing outline; however, when he goes to the word processor, the topic takes on a life of its own. You will see this confirmed in the thoughts of Phillip R. Craig, which follow.

Yancey told me that he has created a database of quotations, more than 6,000 entries, which are cross-referenced to various topics such as doubt, free will, and paradox. He reads extensively, marking quotations for his assistant to enter into the database. He has a favorite quote, "I only read writers who are not only better than I am, and not only better than me, but who write better than I ever will."

## Phillip R. Craig, Writer of Mystery Novels

Phil wrote his first novel, *Gate of Ivory, Gate of Horn*, in 1969. It was twenty years until he had his next novel, *A Beautiful Place To Die*, published at the age of 55. Since 1989 Scribner has published one of his novels every year. Phil's favorites are *Cliff Hanger* and *Vineyard Shadows*, while his wife's favorite is *Off Season*. All of his mystery novels take place on Martha's Vineyard, where he lives full-time.

*Writers don't have lifestyles. They sit in their rooms and write.*

—NORMAN MAILER

PHIL'S TIPS ON BECOMING AN AUTHOR.   Phil believes that if you want to be a writer, read lots of books to learn style and discover ideas. Then, most of all, show up! Go to your word processor and get to work. Writing is indeed hard work, and one shouldn't expect it to be anything else. Sit down and do the best you can.

If you have the opportunity to be around a genius in the field, by all means do so. Learn all you can. Pick his brain. Eventually, you will feel ready to branch out on your own.

When I asked where his ideas come from, he said that the first spark of an idea might be reading in the Boston paper about a man beating up his mother and throwing her out of the window. The body is found on the street. That story was the spark of an idea for a mystery novel. Phil then heads for the word processor and the plot becomes an emerging event, but sometimes he sits at his computer and virtually nothing happens. Other times he gets into a flow and it is like "watching a movie I have never seen" because it just moves and the words come smoothly.

In answer to my two favorite questions, "What is the most satisfying part of your work?" and "What is the least satisfying?" Phil said:

*Most Satisfying:* "I seem to have no choice about writing. It is like a calling. I just have to get it out. Some might call it a curse, but to me it is pleasure to get it out, and look back and see it finished."

*Least Satisfying:* "The least satisfying is the hard days—the days when nothing seems to happen. Writing requires such time demands!"

I asked Phil if he felt his inspiration came from God, and he said, "No." He also felt there was no muse sitting on his shoulder feeding him ideas. Sit down, work hard, get into a flow; that's how it happens according to Phil. At one time, Phil considered becoming a minister, so I was mystified by this comment.

### Charles D. Taylor, Writer of Novels

Charles "Chuck" D. Taylor was my first math editor at Addison-Wesley in 1969. He went on to become a very successful writer of adventure novels in the Tom Clancy genre. Some of his best-sellers are *Show of Force, Boomer, Sightings,* and most recently *Igniter* about an arson investigator in the Boston Fire Department. To my two favorite questions, Chuck gave these answers:

*Least Satisfying:* "I have published fourteen books, and still have three manuscripts that have never been published. Probably the most frustrating thing is the first 15 percent of a book—the Valley of Despair. I do lots of research and grappling with ideas. For *Igniter* I read about serial murderers and psychopaths and spent time with the Boston Fire Department and the Arson Squad. Once I got all that research done, I still had to redirect the book and get started. I never know how it's going to end. I yearn for the confidence in having a sense of what the other 85 percent will be. Once I get that, I can whip out the manuscript in about three months. I have mental blocks and even after fourteen books, still wake up in the night and think I might not get the next book finished."

*Most Satisfying:* "There are a series of satisfying moments after you finish the manuscript and get into page proofs. Before that I enjoy working with a good editor who stimulates my thinking, gives me a new perspective, or even something as small as changing sentences. When finished, I really enjoy autographing and discussing the book with readers. Somehow they think you are God and it is great for your ego. For my military books, the people in those careers really enjoy the books because they will see the pros and cons of their military life. The firefighters were just so happy to have someone write about what it is like to be in a fire. They are so thankful that you are making the public understand what their professions are like."

When asked how to teach somebody how to write a book, Chuck said lots of people ask him that question but he doesn't have even a vague idea about what to tell people. He sees other authors work all kinds of ways. "You have to sit down and get your ideas together. You read *Publisher's Weekly*. You have to get an agent who will help you. Some authors put all their ideas on 3 x 5 cards and assemble them into a book. For me, I do my research and do my composing on the computer. Writing is something that is in you that you can do; it is something you want to do. Lots of people can't get past the Valley of Despair."

On whether writing talent is God-given, Chuck feels it is just a given. Where the original talent comes from, Chuck just doesn't know. He says, "I make my money from creating really bad guys and lovable heroes. I don't know whether that is a God-given talent. You really have to work hard at that so the reader hates those guys and I also work awfully hard so people then identify with the hero. Is creating bad guys a God-given talent?"

I am continually surprised at the books that would not seem to be successful, but are so. Adolph Hitler's *Mein Kampf* was a powerful piece of nonfiction literature. He had a talent to teach people to hate. Was that talent a gift from God? I doubt it! How do you compare Hitler's talent to that of M. Scott Peck? There may be as many answers to the source of authors' talent as there are authors.

Chuck agrees with Peck, Lewis, and Craig that writing is a lust, a passion, a drive to get it out. For his books on the Navy SEALs, the passion was to relate their real story, conveying their intelligence and belief that they can overcome just about anything. What goes on in a firefighter's mind? Why do they walk into burning buildings? What drives them to automatically climb the many floors of the World Trade Center? There is a lust to convey something like that. Chuck says, "When characters are to fall in love, I try to convey as much as I can of what is going on in their minds. In one book there was a Russian admiral who had a wife back home dying of cancer and I tried to convey the very human nature of their love at the time of the Cold War when it was expected that Americans would hate Russians."

### Neil A. Campbell, Author of Biology Textbooks

By far the best-selling college biology textbooks, Neil's titles, published by Benjamin-Cummings, a division of Pearson Education, dominate 60% to 65% of the entire introductory biology market. Dr. Campbell has earned a reputation as an outstanding teacher with a strong commitment to improving undergraduate classroom education.

In answer to my two favorite questions, "What is the most satisfying part of your work?" and "What is the least satisfying?" Neil said:

*Least Satisfying:* "Biology is just exploding in every area, especially in the past 15 years or so. One of the most difficult challenges is to write a book that is manageable and accessible to students, but keeps up with the big progress in biology. From molecules to ecosystems, you want a book that students can understand but that tends as much as possible to all the individual specialties of the biology professors. A common complaint from many professors is that biology books are too big and encyclopedic, but don't have enough on a particular field (usually the professor's specialty). While that conflict is a challenge, it also makes it fun to find the most appropriate balance of breadth and depth."

*Most Satisfying:* "The fun part is continuing to learn across many fields of biology. Indeed, I enjoy talking to other biology specialists. It maintains my knowledge across the broad field of biology. The other reward is as an educator. It feels good to help students learn about what has become a difficult subject. I enjoy the letters and e-mails that come from students, especially graduate students, who keep their books and contact me while in medical or graduate school."

On suggestions on whether a person should be a biology textbook author, Neil would ask some questions: How is your writing? Writing for first-year students is not like writing research papers. How do you think about the subject? Do you see themes that apply across all of biology that can provide touchstones for students who are trying to learn this vast subject? Do you have a tendency to pick up a pencil and make diagrams that help you understand biology—not necessarily a true-life picture, but a diagram that will help students understand some biological process such as how a fish's gills work? Biology is a visual science.

When I asked Neil whether being a good biology writer was a God-given gift, he said that he always liked to write. He continued, "In fact, I began college as a history major partly because I knew I would do better on essay exams than on multiple-choice tests. I don't know how much of good writing is a natural talent and how much of it is learned from practice. Maybe the most innate of the required talents is the ability to make connections, to see how things fit together when others only see the pieces.

Although I always enjoyed writing, I can't say that authoring a biology textbook was a calling. Call it serendipity, or a fortunate discovery made by accident, but Jim Behnke, an editor for Benjamin-Cummings and now President of Addison-Wesley Publishing Company, was visiting my department at Cornell with his boss. My office door was open, and they just popped in to introduce themselves and ask some directions. We ended up chatting so long that they missed their plane. Jim talked me into doing a sample chapter. Once I started, I realized it was something that I was supposed to do."

There are many books on writing in bookstores. A few that I recommend are William Zinser's, *On Writing Well*, Stephen King's *On Writing*, and Michael Card's, *Scribbling in the Sand: Christ and Creativity.* Lots of sources about "creativity" can be found using an Internet search engine.

## THE REVIEWING PROCESS

M. Wiles Keller, in charge of mathematics at the regional campuses of Purdue in cities like Michigan City, Fort Wayne, and Indianapolis, was nearing retirement from Purdue University as I was completing my Ph.D. He interviewed me for a position at the Purdue Campus in Michigan City, Indiana. Interestingly, he was also an author and had a favorite saying that Keedy often quoted: "Good books are not written, they are rewritten!"

An author must be resigned at the outset to many drafts or rewrites of a manuscript. It is a tendency, which is resistible as an author becomes successful, to think that what he or she writes the first time is cast in stone and needs no refinement.

**REFERENCES**

Zinser, W., *On Writing Well.* New York, New York, Quill A Harper Resource Book, 2001.

King, S., *On Writing.* New York, New York, Pocket Books a Division of Simon & Schuster, Inc., 2000.

Card, M., *Scribbling in the Sand: Christ and Creativity.* Downer's Grove, Illinois, InterVarsity Press, 2002.

*An absolutely necessary part of a writer's equipment, almost as necessary as talent, is the ability to stand up under punishment, both the punishment the world hands out and the punishment he inflicts on himself.*

—IRWIN SHAW

A good textbook has undergone a series of possible rewrites:

1. By the author to polish the manuscript
2. By co-authors, if any, as they critique the writing and blend writing styles
3. By reviewers of various drafts of the manuscript
4. By manuscript accuracy checkers
5. By copy editors after the "final draft" of the manuscript has gone into production

When a series of books like the Trilogy is so successful, it is typical to get ten reviews before a revision is started. The editors and authors use these to make decisions about what to do in the next edition. Then a revision plan is created and the author goes to work. Upon completion of a first draft, the manuscript is read by co-authors, and the real "first draft" is completed and submitted to the publisher, who then may obtain ten more reviews from which a final draft is created.

Even after a "final draft" is created, a publisher may obtain even more "marketing" reviews, which editors use to determine key aspects of the books as sales tools. I used to read such reviews and feel compelled to polish the manuscript again. In more recent years, I just have the editors inform me if any math errors need to be corrected. But one of my co-authors deals passionately with such reviews. I know he and other authors who even revise or rewrite books after they have been typeset. Publishers hate this because of the excess costs, not to mention delay in getting the product to market in a timely way, and I avoid it if at all possible. One might ask, amid all this rewriting, just what is a "final draft?" I am not sure I have a pure answer.

The most satisfying part of my profession as an author is finding out how well students have learned from my textbooks. This can come directly from students, but most often from instructors using our books. The least satisfying part is dealing with criticism, most of which is sought by the publisher from instructors through reviews. There is a paradox with reviews: Ignoring them will eventually lead your books to failure, yet responding to all criticism is impossible and quite often redundant. I would also add that one's pride often gets in the way. I continually resist the thought, "I have written successfully for thirty plus years, how can anybody know more about writing a textbook than I?" The day you stop paying attention to reviews is the day you should retire from writing.

Suffice it to say that anyone who dislikes rewriting and being criticized by peers should not become an author!

It also bears commenting that an author can have too many reviews. Too many reviews can offer too many comments that might turn creativity into "vanilla."

M. Scott Peck writes regarding reviews, "I can guarantee in advance that any new book of mine will receive at least one rave review and one that will

*Proverbs 15:22: Plans go wrong for lack of advice; many counselors bring success.*

*Proverbs 29:1: Whoever stubbornly refuses to accept criticism will suddenly be broken beyond repair.*

REFERENCE
Peck, M. S., *In Search of Stones.*
New York, Hyperion, 1995,
p. 349.

*To write a successful textbook,
you strive for that level of
mediocrity that makes for
success.*

—M. WILES KELLER

disparage the work as excrement, and it will be hard to imagine that the reviewers read the same volume."

Reviews can take an author in a number of directions. It's important to look for trends in reviews. In theory, I try to respond to every line of a review, although reviews often conflict with one another.

Reviews can actually spur creativity. Often the same complaint crops up in several reviews. In the process of grappling with it, I might create a resolution never once suggested in the reviews, and the book is better as a result of that prompted creative solution.

## The Paradox of Rigor vs. Low Level

Textbooks are bought by students after they have been chosen by the instructors through an adoption decision. An author might write a rigorous book that appeals to the mathematical instincts of an instructor but the students can't follow. Conversely, the most perfectly readable book for students may never reach them because the instructor deems it too low level.

Perception is an interesting aspect of the publishing business. We once wrote an algebra-trig book much like our successful developmental series, the Trilogy, and it was published in paperback. But, the market perceives a paperback textbook as easy or low level while a hardback book is considered more rigorous or high level. To combat this reaction, we took the paperback and, in effect, cut out the margin exercises, redesigned it, and formed a hardback. I had an editor from another company insist that the hardback was more rigorous than the paperback—but it was the same book!

## Letters from Students and Reviewers

Let me show both sides of the feedback an author receives with the following negative comments and then positive sets of comments from students and/or reviewers.

ON THE NEGATIVE SIDE

*Mr. Bittinger, you may know math but you cannot teach it. Your book is useless in trying to understand algebra. You should know, as I have looked through many others and all are more useful. I think that you are just a math geek idiot! Thank you, student of math despite you.*

—A student

*I have been teaching college for twenty-five years. During these years, some of our math faculty have always used Bittinger texts. I have always felt that you think like a university professor trying to simplify down to the level of the community college students who have always had difficulty with math. You don't know how nonmathematical students perceive math ideas. You have always presented material in somewhat of a cookbook manner, which encourages students to memorize, rather than understand. My feeling is that you were good at mathematics and really don't understand why people would have any difficulty with the subject. I can see over the years, you have mentally accepted that not all students are as bright as you are, but I don't feel that you emotionally relate. I know your texts have been popular with instructors for a long time, but maybe it is because you just give rules and practice problems that are easy for teachers to use.*

—A college math instructor and reviewer

In the latter case, I wonder if the reviewer would be interested to know about that D I received in second-year algebra because I rushed and made silly mistakes on tests, or that I was on academic probation after my first quarter in the master's program at Ohio State.

ON THE POSITIVE SIDE

> *Writers become idiotic under flattery sooner than any other set of people in the world.*
> —FRANK MOORE COLBY

*Once again Bittinger has lived up to his reputation as an outstanding mathematics textbook author. He has prepared an outstanding manuscript and probably one that I will adopt when it becomes available. Mr. Bittinger has an uncanny ability to display the mathematics in such a way that all students can enjoy and understand the material. I am very appreciative of his efforts. His work will benefit my students tremendously.*

—A community college instructor and reviewer

*I have just started some higher math courses past your last book,* College Algebra and Trigonometry. *I must say that the entire time I used your book from* Basic Mathematics *all the way up to* College Algebra and Trigonometry, *I did not once require assistance from my professors. Now that I am taking beginning calculus, I am having trouble understanding the book from page 1. My newest professor said that your books are the easiest of all math books to learn from and that he did his undergraduate studies using the same books, but that you don't have books available at the higher levels. Is this true? I sure hope not. I am not looking forward to pouring over this calculus book for another six hours. If you do have any higher-level math books or if there are any in the planning, then please let me know.*

—A student

The following is my all-time favorite.

*Thank You! In order to enter a nursing program that I've saved for years to enter, I found out that I had to pass a math test. (I probably should have checked into that before quitting the job, but then it was too late.) "You have to have some basic algebra concepts" they said. I knew I wouldn't do too well, having flunked algebra twice in high school before passing through pity, but I took the test and scored a 22%. "You're going to need a year's worth of math classes before you can enter," they told me. I asked them if I could take it again, but they said I could only take it one more time, and I'd never be able to pass it in seven weeks. Passing was over 85%.*

*Not having an extra year's worth of rent, I was determined to learn algebra in seven weeks. I bought an algebra book that proposed to be quick and easy, and tried that. Ha! Ha! I then realized that there is no easy way, and checked out all the books at Borders. Yours looked the best, so I bought it. Pressured and upset, I sat with that book all day every day. (There is no sweat in the book, but I had a paper cut I didn't know about, so there is blood, and there are also a few tear stains.) I didn't go out, didn't visit with friends. You were my only friend for seven weeks, and you did more for me than most people I know. Sometimes I would look at your picture and think, "He looks like a nice man. He is giving me a good book that will teach me what I need to know. I just have to keep going."*

*Thanks to your fantastic book, I was able to LEARN algebra all by myself. I got a 96% on the test! The testing woman, before I went in, even said, "It's only been six weeks since you tested. You understand that if you score lower, you may have to drop down a class, don't you? Are you sure you want to do this?" When I came out and she scored it, she asked me if I'd like to take the next higher test!! I was in. My life can move forward. I will get to be a nurse instead of working in a dumpy, cold office in a warehouse. Your book has made my life better!!*

*I signed up for an Intermediate Algebra class at school, but the book was just copies of other books, with all the explanations left out, bound together. The teacher wasn't too hot, either. I dropped the class and bought your Intermediate Algebra book. I hope to be ready for College Algebra when I'm done.*

*Marv, even though I got very mad at you when I had to factor all those different nomials, and drew martian ears on your picture, I just had to say thank you! Not only is my life now moving in the direction I wanted it to, but after all these years I have gained the satisfaction of knowing that I am not too stupid to learn algebra. I realize this may all seem a little silly, but I truly am so grateful that this book existed. You needn't respond to this e-mail. I just had to write it.*

*THANK YOU! THANK YOU! THANK YOU!*

*—A returning student*

*P.S. Why, why, no answers for even problems? Very often I do a problem and get it wrong, then go back and do it over. Sometimes over and over and over. I see no point to doing it wrong and not knowing, so I can't use the even ones, and I lose half the practice! But that's OK. You're still great.*

This latter note fulfilled my dream about why I write mathematics textbooks. I keep the five of these notes posted on my bulletin board to keep life in perspective. "I live in paradox!"

A humorous story about noted math textbook author Earl W. Swokowski offers some insight about textbook reviews. I knew Earl mostly by reputation as a fine author and gentleman. His first textbooks were a series of successful precalculus books before he went on to write a best-selling calculus book. When Earl wrote a book, students and professors could trust that the mathematics was correct and well structured.

Earl didn't attend many meetings and seemed not to like appearing before groups. Only once did I have the pleasure of talking to him and hearing him speak at a mathematics convention. He was very forthright in discussing his experiences writing textbooks. What I remember most was a story he told:

> *I recall revising my calculus book once, and struggling with the reviews from a particular university with a large adoption. They said in effect that "If you do not do everything we list in this review, we will drop your book!" Earl then told us with a wry smile on his face, "Well, this time I did everything they said. Then they dropped the book anyway."*

I roared when I heard this, because it went right to the core of my experience as an author—reviews can be challenging, and controversial, and they sometimes even violate the author's educational instincts.

## EDITORS, ROYALTY ADVANCES, AND USED BOOKS

Anyone who decides to write a textbook eventually encounters editors of many types. At the top of the hierarchy is the Editor-in-Chief, sometimes called Publisher, who is in charge of all the editors in a specific field such as mathematics, or in two or three fields such as math, computer science, and statistics. An author tends to have the least amount of day-to-day contact with this type of editor, who has management duties in the company.

At the next level are acquisition editors, who acquire manuscripts, negotiate contracts, and brainstorm with authors about the creation of their books. Acquisition editors receive a bonus every time they sign a contract.

When developmental editors work on a book, they analyze a manuscript in comparison with its competition and often take reviewers' comments into consideration. They might spot a missing topic, or a way to improve the design, or a flaw pointed out by the reviewers.

Then there are project editors who find reviewers, watch deadlines, obtain permissions, and even work on supplements to the text.

When a book goes into production, it first gets a word-for-word, line-by-line editing by a copyeditor. These people labor tirelessly behind the scenes and deserve so much praise for the end quality of a book, yet they often fail to receive credit for their role in the complex process of book production.

The first step for an author after a manuscript is released to production is reviewing a "copy-edited manuscript." While it may seem like thousands of marks have been inserted for changes, you must not take offense. I refer to them as "red spaghetti-filled pages" because the tome you created to save the world seems obliterated. This part of the publishing process is as normal as the sun coming up each morning. It is just one more detailed rewrite that you did not expect. Keep in mind that someone with a great deal of talent directed in another dimension has polished your manuscript. By all means, check all the marks, but keep in mind that you have a better product as a result of this editing.

Noted author M. Scott Peck writes as follows about the first time he received one of these marked-up manuscripts.

REFERENCE
Peck, M. S., *In Search of Stones*. Hyperion, New York, 1995, pp. 234–235.

*So, I'm grateful to God for my books. But I am also grateful to editors. My . . . reaction, . . . prolonged, was one of rage that any human being could possibly be so nit-picking. Only after twenty-four hours did I come to a set of more balanced realizations. One was her suggested minute changes, with very few exceptions, made it a better book. Another—and this was crucial—was that she and I had different personalities and hence different gifts. Had I possessed her extraordinary attention to detail I probably wouldn't have been able to write the book in the first place. On the other hand, if she possessed my capacity for "big picture" thinking, she probably wouldn't have been able to be a copyeditor. I needed her nit-picking, and she needed my book to pick on. We needed each other. Her gift of nit-picking was, in fact, a gift of love, and ever since I have been profoundly grateful for the service of copyeditors. They may have been regarded as being at the bottom of the heap, but for me they have been partners.*

This quote leads me to recall once receiving a red-spaghetti-marked manuscript from my co-author, Mike Keedy. He calmed my distress by informing me that if the material had been bad, there would be little to mark up. But if it is good, then it can be polished and honed to be even better. Remember the quote cited earlier, by M. Wiles Keller, "Good books are not written, they are rewritten!"

Most of an author's day-to-day work is with an acquisition editor. I have worked with more than forty editors over the thirty-plus years I've been writing. The first was Charles D. Taylor, whom I mentioned elsewhere in this book. He and Greg Tobin, mentioned in the introduction, are the only ones yet named.

I have had some terrific, creative, loyal, and hardworking editors over the years. Those that were the most effective and most memorable were those with whom I was able to build a personal relationship. I felt like I was working for them rather than for Addison-Wesley. I thank them all and miss the ones who by nature of the business world have seen fit to move on. The authors stay, but the editors move up or out.

One last point on editors: Ignore their college major. The good qualities of an editor evolve from their abilities at appraising a market, making creative suggestions, and being encouraging. I had only one editor in thirty plus years who was a math major, and he was a poor editor who brought too many of his own math biases to the forefront.

## Royalty Advances

Sometimes an author can negotiate an advance on anticipated royalties with a publisher. An advance can be especially helpful when a textbook author is first getting started and needs release time from summer teaching or maybe even a semester off. Advances are paid back to the publisher from royalties once they start coming in. I tend to avoid advances because I don't like to get my reward before I do the work. But, if one author requests an advance, then the co-author(s) should take one as well in proportion to contractual royalty split. Then all are protected if the book is not published. (Typically, if a book isn't published, authors are required to repay the advance.) I once co-authored a book where my co-author took a huge advance. I worked hard on that book, but it fell through and was not published by Addison-Wesley. He got his advance, but I got nothing.

### Used Books

I am often asked, "How do you revise a textbook? Isn't math the same today as 10 years ago?" The math might remain the same but there are trends in math education that authors respond to. Examples in the early parts of my writing were the metric system, ethnic and gender issues, and the scientific calculator. I recall actually counting the number of female and male names that occurred in our books as a response to the gender issue, and also writing problems that involved no gender at all. More recent trends we respond to are the inclusion of applications and the graphing calculator. Responding to these trends was a major challenge in our revisions and demanded a good deal of time and energy.

But, what is the real, honest, definitive reason to revise a book? To respond to the destructive effect of the used-book market! I could write an entire book about this issue but will try to limit myself to a few comments. Typically, if a book sells so many copies one year, it will sell about two-thirds of that number the following year, and the trend will continue until the book is revised four years later. By the third and fourth years, sales have decreased tremendously from that of the first year.

I have little to complain about if there is some kind of student-book exchange, where a student can sell a used book at a fair price, and another student can go there and buy it used at a fair price. But, when used-book companies come to a campus to purchase used books, they don't pay a fair price and the books are typically marked up four or five times before being sold again at another campus. This gets compounded tremendously by the used-book people who buy "comp" copies given free to professors to consider for adoption. The professor sells them to used-book people and makes what seems like an easy profit. I call these kinds of used-book people "the vermin that roam the halls!"

There is a classic story told about a book written by two Addison-Wesley authors. The only adoption of the book was at their own college, and it was large, 3,000 books. Do you know how many new books were sold at that college in the fall? The answer is the empty set. The entire adoption was filled with "used" books. The only place those books could have come from was the purchase of comp copies from professors. How would you like to spend three years of your life writing a textbook and get no compensation?

*Instructors can't complain to publishers about the price of new books to students when they continue to sell comp copies to the used-book vermin!*

I am well aware and concerned that the price of new books has sky-rocketed over the years. There are reasons beyond the effect of the used-book market: (1) the cost of paper, (2) the demand for four-color books, and especially (3) the demand for supplements such as videotapes, instructional software, and computerized test banks. Instructors demand these to be give-away items, but they have to be paid for out of the price of new books.

I wish the prices could be lowered. In 1971 the price of my *Intermediate Algebra* paperback was $7.95. In 2003 I found the same book in a bookstore priced at $95. When I give talks to instructors, no topic becomes more emotional than the price of books. Several years ago Addison-Wesley did a financial study of the effect of used books on the publishing industry. They found that if students restricted their purchases to new books *only*, then the price of a new book could be lowered by 30 percent, and the bottom line (profit–loss) to the publisher would be the same.

Let's make a comparison with the music industry. If you write a piece of music and I play it somewhere for money, then I am required by law to pay you a royalty on what I made for my performance. Ironically, if a used-book company sells a book, it has no obligation to pay a royalty of any kind to either the publisher or the author. This is not fair. The used-book company uses my creative endeavor, published at a substantial cost to the publisher, yet pays nothing for the investment of time and money. Without a change in the copyright laws regarding publishing, this curse on the publishing industry will persevere.

# On College
# Mathematics
# Education

# 7

Among the many insights that I learned from my co-author and mentor, Mike Keedy, was that educational history seems to have a bad habit of repeating itself. Educators start with a great idea, a *baby*. The baby is cute and has its merits, so much so that the educational community gets fired up about the idea and climbs on a *bandwagon*, adopting the idea whole-hog and wringing it for all it is worth. Inevitably, as with any good idea, negative aspects occur because of the hurried way it was implemented. Then the world decides, perhaps erroneously, that the idea is totally bad, and throws it out entirely with the *bath water*. In the process, the good parts of the idea get discarded as well.

A prime example was the modern math movement of the 60s and early 70s. It was motivated by the Russian satellite *Sputnik* in 1957, the first in orbit. The idea of advancing scientific training in the United States was the baby. The bandwagon was the jump to carry it out by pushing advanced concepts down to lower grades. The bath water was throwing out the entire idea because it downplayed the teaching of skills. Just because some aspect of nondecimal bases can be taught to third graders does not mean it should be, especially if skills get slighted and teachers are ill prepared to teach it. Just because some aspects of calculus might be taught in ninth-grade algebra does not mean they should be, especially if equation-solving and problem-solving skills get slighted.

One aspect of the modern math movement was worth saving and that was the adherence to providing a rationale to the mathematics being taught. In my opinion, providing the rationale got thrown out with the bath water. The rationale of concepts should be provided *without sacrificing skills*. Thirty years later, I still ascribe to that in my writing philosophy—it is in all my books! I try to write with understanding, but my goal is sound mathematics skills as well. Where does this understanding come from? The answer is by experimentation or discovery, intuition, logical reasoning, and finally abstraction.

## THE FUNDAMENTAL THEOREM OF MATHEMATICS EDUCATION

There is a better approach to new educational ideas. Suppose when the notion of mastery testing came out a few years ago that a couple of people in a math department were excited about the idea, say in intermediate algebra, even though the rest of the department was either disinterested or neutral. Suppose further that the teaching of a mastery-based intermediate algebra course was not mandated for the entire department, and those two interested instructors were allowed to try it out in two experimental sections.

In such an experiment, the two instructors would be able to confer with each other as they taught, and they would inevitably find methods of teaching that worked and others that did not. Then they could refine the course and teach one more semester experimentally before going back to the department with recommendations for renovating the entire course. If mastery testing did have positive results, their revised methods of carrying it out might prevent it from being lost in the bath water because it was forced on the department without proper evaluation.

Professor Judith A. Gersting and I conducted such a mastery testing experiment at IUPUI some years ago. The rules we mandated in our two classes were that if students did not get a 75 percent on an exam, then it was necessary to retake the exam until achieving that goal. If they did not achieve their goal after five chances, they had two choices—receive an automatic F in the course or withdraw and go to the prerequisite course. Needless to say, all that testing took lots of time and resources.

Our results were that the same grade distribution was achieved with the mastery testing as without. That is, there were the same percentages of A's, B's, and so on. In that sense, the results did not warrant the extra work. We did achieve an advantage: Students found out sooner that they did not have the prerequisites for the course and dropped back. In the normal course, they might have delayed going back, to their detriment. Our recommendation to the department was not to implement mastery testing because of the demands of time on ourselves and our teaching assistants.

My premise regarding educational change follows.

### The Fundamental Theorem of Mathematics Education

To implement educational change,
1. start small with people who are enthusiastic about the change,
2. let them experiment with the change, finding good parts and bad parts, and
3. then make recommendations for further implementation.
4. Do not implement any educational change that neglects skills. Doing so will fail your students in both the short and the long term.

### Calculus Reform Movement

Recently, the word *reform* had its jump-start in college publishing through the recent reform in teaching calculus. The textbooks published to reflect this reform experienced a boom in sales because the publishers responded to the trends. I hate to be an "I-told-you-so" guy but, as soon as I heard of those courses, I realized the skills had been slighted and I was certain the students would suffer. Sure enough, when students got to the third semester of calculus, their poor skills let them down and they struggled. Calculus reform is now on the decline. I see no reason that calculus can't be taught as a union of skills *and* concepts!

I would go so far as to say that *any* reform movement that de-emphasizes skills is doomed to fail its students. The Fundamental Theorem of Mathematics Education is so insightful in analyzing education trends that it warrants being an  .

 MUSIC: "The Ride of the Valkyries," from the opera Die Walkure by Richard Wagner, track 3 on the CD *Greatest Hits Wagner*, SONY Classical.

## GRAPHING CALCULATORS AND DISTANCE LEARNING

Let's apply the Fundamental Theorem of Mathematics Education (FTME) to two recent trends in math education.

### Graphing Calculators

The appeal of this technology to instructors is comparable to the appeal of a new video game for an adult. The power of new technology is captivating. In the case of graphing calculators (graphers), the appeal lies in the ability to quickly visualize mathematics by seeing graphs. The other attraction is the quickness of doing other complicated processes like solving equations and regression. I am quite enamored with regression because it provides a succinct way to get at the eternal question students have, "What is this good for?" The instructor can bring tables of data from newspapers to the classroom and do regression, and make interpolations and extrapolations, thereby silencing the students' eternal question. Other advantages might include the use of experimentation, developing the notion of programming, and the use of logic.

I applied the FTME when I conducted a graphing calculator experiment at IUPUI in 1996, when we were writing a graphing-calculator algebra-trig book. Following are the details and results of the experiment.

### The Experiment

The experiment was done with 55 students in a college algebra class in the spring of 1996. Students were required to "lease" a TI-82 graphing calculator.

REFERENCE
Bittinger, M. L., Beecher, J., Ellenbogen, D., and Penna, J., *Algebra and Trigonometry: Graphs and Models*. Boston, Pearson/Addison-Wesley, 2001. There are related precalulus, college algebra, and trigonometry titles.

They got their money back at the end of the course if they didn't want to keep the grapher. I bought all those calculators, gambling that the students would fall in love with them and not want to sell them back. My gamble worked since only 30 percent wanted their money back and it was easy to sell the remaining calculators to other students.

One of the constraints of my being allowed to teach this experimental course was that the students had to take the departmental final, but they could not use their graphing calculators. This almost demanded of me that I not slight the algebra skills. My goal was to teach a course that balanced graphing-calculator and algebraic skills.

My plan was to use the graphing calculator in the class in much the same way as we now use a scientific calculator; that is, make use of the graphing calculator as an adjunct to an algebraic class.

I taught the following basic graphing calculator skills before starting the normal course:

- solving equations
- finding points of intersection of two curves,
- understanding trace and zoom,
- using the table feature,
- making graphs, and
- working with identities.

I was most enthusiastic about identities and regression. How many times do students ask why isn't $\sqrt{x^2 + 16} = x + 4$? The quick response now can be: Check the graphs. Check the tables. For example,

| Equation | Is it an identity? |
|---|---|
| $\sqrt{x^2 + 16} = x + 4$ | No |
| $\dfrac{x^2 - 9}{x - 3} = x + 3, \quad x \neq 3$ | Yes |
| $\dfrac{\pi}{2} - \dfrac{4}{\pi}\left(\cos x + \dfrac{\cos 3x}{9} + \dfrac{\cos 5x}{25}\right) =$ $\dfrac{\pi}{2} - \dfrac{4}{\pi}\left(\cos x + \dfrac{\cos 3x}{9} + \dfrac{\cos 5x}{25} + \dfrac{\cos 7x}{49}\right)$ | No |

Regression was done as it came up in the course with linear equations, and later with polynomials and exponential, logarithmic, and logistic functions.

## Results of the Experiment

I passed out a student survey at the end of this section. All the positively directed questions averaged 3.00 or higher on a scale of 0 to 4.

The students took a departmentally written and administered final examination on which they could not use their graphing calculators. A statistical study was done to compare the graphing-calculator experimental class with the rest of the department's students who were taught a traditional, nongraphing-calculator course. The statistics are summarized as follows.

*Graphing-Calculator Class*

Sample Size = 28,

$\mu_1 = 60.68$,   $\sigma = 19.74$

*Other Classes*

Sample Size = 283,

$\mu_2 = 57.27$,   $\sigma = 22.85$

$T = 0.76$,   $p = 0.44$

The null hypothesis $\mu_1 = \mu_2$ was accepted via a statistical test of ANOVA. The differences in the means were *not* statistically significant, but the mean of the graphing-calculator was higher than the nongrapher class. I felt safe in drawing the conclusion that a graphing-calculator class can be taught in a way that does not diminish skills.

## Other Observations

- There was only one instructor, myself. Instructors had not been placed randomly in the experiment. Would the same result have occurred anyway?

- Students were not placed randomly in the grapher class. A sign in the bookstore regarding the grapher requirement explained that they could select the course.

- Of the 54 students who were required to buy a grapher, only 16 returned the grapher to get their money back. That is, 70 percent bought the grapher and kept it.

- It would be worthwhile to track these students into trig and calculus for further decision making.

- The experiment should be run again with more instructors in the grapher section.

## Recommendations and Suggestions

- Students need exercises each day that support the grapher. The class was using a text that was not grapher driven, so they felt the grapher got slighted at home. They had to decide to use it most of the time. On the other hand, this may be the reason they survived the final.

- Math departments should require the graphing calculator in algebra-trig. The use of the grapher is sweeping the high schools. More and more students will be coming to us with grapher background and asking why it can't be used. Virtually all textbooks you consider in the future will be highly grapher driven, although the grapher might be optional.

- Keep algebra skills in the forefront of the course—maybe a balance of 75% algebra, 25% grapher. But use the grapher to speed up and enhance overall visualization and topical coverage such as rational functions, polynomials, and even matrices. If you minimize algebraic skills too much, students will have a disaster in calculus.

- Be sure to use an overhead view screen in class. Have the screen off to the side so the blackboard or other overhead is still useful.

- Each class session, have a different student do the keyboarding on the grapher that drives the view screen. This frees the instructor to concentrate more on the lecture and to move freely about the room. This also gives the instructor a chance to see how various students are doing. This is a very useful tip that I picked up at a math convention.

- Require the use of the TI-83+ (new) grapher of all students in alg-trig. The TI-83+ allows the drawing of several graphs in such a way that each graph looks different and has new regression features to fit logistic and trig curves to data. The TI-83+ is also more matrix friendly (all operations preprogrammed) and has better statistical capabilities, such as ANOVA.

- An extra teaching assistant who is grapher literate would be very helpful the first day or two of class.

- Another issue the instructor needs to tend to is the inclusion of some questions on exams that give feedback as to whether the student is learning to use the graphing calculator.

- Obtain posters of the keyboard from Texas Instruments and use them early in the course to show the keystrokes to students.

- I would definitely teach a grapher course again with great anticipation, but I would insist that students be allowed to use the grapher on the final. Not allowing the grapher created too much anxiety. It did, of course, allow for a revealing statistical comparison.

### Distance Learning

Distance learning fills me with concern. It is another "Babes, Bandwagons, and Bath Water," disaster waiting to happen. Distance learning may have been started by college administrators eager to obtain funds for innovative programs. On first consideration, the idea of reaching students unable to attend a college seems a worthy goal. I think it will last only as long as the government and/or universities are willing to invest in it. After the funds dry up, the baby will be thrown out with the bath water.

I once heard a talk at an AMATYC (American Mathematics Association of Two-Year Colleges) meeting by an excellent and hard-working instructor. She was teaching a small distance-learning class but expending incredible resources trying to reach her students by faxes, e-mail, or phone calls. Is it reasonable to think that such an expenditure of time and resources can be extended throughout developmental teaching at the college level? This was a dedicated, hard-working instructor who believed in what she was doing. Give this task to an uninspired instructor and it will die on the vine.

I often visit community colleges when they are discussing the use of my books. Once while I was waiting in a hall for instructors to arrive, two deans were sitting in an adjoining office chatting with great enthusiasm about their distance-learning course. After some time, I interrupted and asked how many students were in the course. The answer was twelve. Twelve students and all that work. How utterly impractical!

Dr. David M. Matthews, Vice President of Instruction at Southwestern Michigan University, during a talk at Addison-Wesley's annual ICTCM conference, gave five compelling reasons that universities should not invest money in distance learning:

1. Instructors can just as well use the technology in existing courses.
2. It requires too much time of both the students and the instructors.
3. It exacerbates problems of assessment and cheating, as well as the denial of the existence of cheating on the part of the instructor.
4. DL courses may actually widen the learning gap for lower-socioeconomic students.
5. Resources, resources, resources. The time of full-time faculty members is a precious resource. There was not enough of this resource before DL occurred.

REFERENCE
Matthews, D. M.,"The Case Against Distance Education," presented at International Conference on Technology in Collegiate Mathematics, November 2000.

To me issue 4 is of primary concern. The problem with distance learning is *distance*. I have a strong concern about any teaching method that eliminates the "warm hand on the shoulder" provided by direct one-on-one contact with the student.

In a National Education poll it was determined that "faculty believe Web-based courses do a better job of giving students access to information, helping them master the subject matter, and in addressing a varsity of learning styles. Faculty believe in existing courses which strengthened group problem-solving skills, verbal skills, and oral presentations."

The latter is a way to take the good parts from the idea of online distance learning and apply them to existing courses. I hope that this is the way online learning survives.

## EMOTIONAL INTELLIGENCE: A NEW TREND

I once read the best-selling psychology book *Emotional Intelligence* by Daniel Goleman with much fascination. What struck a chord with me was

*The difference between a good teacher and a bad teacher is 18 inches—the distance from the heart to the brain.*
—ANNE GRAHAM LOTTS

REFERENCES
Goleman, D., *Emotional Intelligence.* New York, Bantam Books, 1995.
Goleman, D., *Working with Emotional Intelligence.* Bantam Books, New York, 1998.

connecting Goleman's ideas with education, even though he never explicitly discusses the topic. For years the trend I would most like to spread throughout the educational world has been an emphasis on people skills, or *emotional intelligence.*

I went into math education to a great extent because I had such a remarkable string of excellent math teachers. Each had a way of figuratively placing the warm hand of encouragement on my shoulder. In short, their people skills were as excellent as their teaching skills.

Goleman explains in his book that, "In a sense we have two brains, two minds—and two different kinds of intelligence: rational (IQ) and emotional (EQ). How we do in life is determined by both—it is not just IQ." (p. 28) This message has had a particularly positive effect on the business world, so much so that Goleman wrote a second book, *Working with Emotional Intelligence,* just for business people in which he asserts, "The rules of work are changing. We are being judged by a new yardstick: not just how smart we are, or by our training and expertise, but also by how we handle ourselves and each other." (p. 3) And, "IQ takes second position to emotional intelligence in determining outstanding job performance." (p. 5) Dennis Wells, a business executive with Arby's restaurants, once told me he never has a meeting with his managers without discussing some aspect of emotional intelligence.

Whether we like it or not, there is a business side to education. Think of the budgets of math departments and how they are affected by large enrollments. Emotional intelligence (EQ) may be even more important in education than in the business world.

### Describing Emotional Intelligence

Mathematicians like well-defined concepts. Since the notion of EQ is not precisely defined, let's try to develop it from many angles. Goleman describes EQ as relating to these characteristics:

- Caring attitude
- Feeling quality
- Love
- Good character
- People skills

Consider the Scouts Law: "A scout is trustworthy, loyal, helpful, friendly, obedient, cheerful, thrifty, brave, clean, and reverent." These characteristics go far to describe EQ, and a teacher with those traits would be far along the road to excellence.

If you are of the Christian faith, the so-called fruits of the spirit are listed in Galatians 5: 22–23: "But when the Holy Spirit controls our lives, he will produce this kind of fruit in us: love, joy, peace, patience, kindness, goodness, faithfulness, gentleness, and self-control . . ." (New Living Translation) These characteristics again go far to describe EQ.

REFERENCE
Simmons, S., Simmons J. C., *Measuring Emotional Intelligence.* Arlington, Texas, Summit Publishing Group, 1997. www.summitbooks.com

A book by Simmons and Simmons describes EQ in terms of character using the adjectives "optimism, self-esteem, commitment to work, attention to detail, desire for change, courage, self-direction, assertiveness, tolerance, consideration for others, and sociability."

Because people define good character differently, some ambiguity may arise in defining EQ exactly. Still, the above terms go a long way in building the concept of EQ, and I think any teacher who could improve in any of these areas would improve his or her EQ.

*I make a special point in every class session to let students see my passion for the subject [of mathematics]. I let students know I care about their success and that I am willing to provide them every opportunity to succeed.*

—PHIL DEMAROIS, AMATYC Teaching Award Winner, quoted in "Opportunities for Excellence: Professionalism and the Two-Year College Mathematics Faculty," AMATYC, 2001, p. 10.

According to Goleman, studies have shown that everyone's EQ can be improved. Another study cited by Goleman analyzed 80 Ph.D. students in terms of success in their fields. The EQ was found to be four times more successful in predicting their success than was the IQ. That is a staggering result!

What do movies like *Mr. Holland's Opus*, *Rudy*, *It's a Wonderful Life*, *The Emperor's Club*, and *Patch Adams* have in common? Yes, they exemplify people with high EQs. I give a talk on emotional intelligence at math conventions. I was so touched by *Mr. Holland's Opus* as it applies to math teaching that it merits an **MPX**.

**MPX** MOVIE: *Mr. Holland's Opus* (1995), starring Richard Dreyfuss and Glenne Headly. To me this movie is what teaching is all about. Poor Gertude Land (Alicia Witt) is trying so hard to learn the clarinet that she forgets to put her feelings into her music. Mr. Holland (Dreyfus) inspires her to embrace her feelings by inspiring her to "see the sunset." Try this in your math class if you are an instructor. Ms. Lang returns at the end for Mr. Holland's retirement, only now as governor, and helps perform Mr. Holland's now completed opus. This is what teaching is all about.

People from the sports world with high EQs include Dusty Baker, manager of the Chicago Cubs; Bill Walsh, former coach of the San Francisco 49ers; Mike Krzysewski, basketball coach at Duke University; and Phil Jackson, coach of the Los Angeles Lakers. Rod Beck, who played for Dusty Baker, once said of him, "If you can't play for Dusty Baker, you can't play for anybody." His people skills are outstanding. Christ displayed an exceptionally high EQ. He didn't beat people over the head about what he wanted them to do. He did it subtly to get their attention. Individuals with a high EQ are concerned about how they are doing with other people, and whether their people skills can be improved.

If you are looking for people who exemplify low or no EQ, Adolph Hitler, Saddam Hussein, and Osama Ben Laden quickly come to mind. Think of the worst teacher you ever had. How was his or her EQ?

## Emotional Intelligence at Work in Mathematics Education

John Beane, who operates a company in China Grove, North Carolina, called Leadership Concepts has more than 15 years' experience with the

*John Beane, Leadership Concepts, P. O. Box 700, China Grove, NC 28023-0778*

Simmons EQ testing techniques. Beane runs seminars on developing emotional intelligence. In a lengthy interview he discussed many issues of EQ pertaining to math professors:

- EQ would help you explore why you do something.
- A math professor might use EQ as a tool to dissect students' needs to learn what teaching approach or method is best for them instead of, "here is my way, take it as it is."
- Great math professors use EQ.
- A professor adept in the use of EQ might be better prepared to know why math appeals to some people and not to others. Beane asserts that in today's society more and more students are lacking EQ.
- A professor whose EQ skills are high might seek a better way to reach students in a math course. He might pay attention to the quiet, seemingly inattentive person hiding in the back.
- The basis of EQ is understanding yourself so you can apply EQ and interact more effectively with people.

Given that EQ is so important in the business world, I would like to see

1. teachers at every level in every field have at least one, maybe two, courses in EQ as part of their training and
2. every student have at least one course in EQ as part of undergraduate education.

REFERENCE

Neptune, C., *Opportunities for Excellence: Professionalism and the Two-Year College Mathematics Faculty*. Memphis, Tennessee, AMATYC, 2001, pp. 22–27.

It is quite common to find articles in newspapers regarding outstanding teachers. What do all of these teachers have in common? You guessed it—high EQ. Instead of spending educational dollars on technology and other quick fixes, let's begin a new trend using the FTME to enhance the EQ of teachers and students. I'd like to see an increase of talks at professional meetings on EQ and have it become a hot ticket for the twenty-first century. Carolyn Neptune has written an excellent booklet promoting teaching excellence for AMATYC. It includes topics on mutual respect, caring, and enthusiasm, all of which exemplify a high EQ.

*It was finals week. I was up to my ears in exams and projects to be graded. A student appeared at my office door and wanted a word with me. Not one of the vocal students in class, not an "A" student, he was like hundreds of students we all have in class every year, so his comment was rather unexpected. He said, "I think sometimes you don't realize what a big difference you make in students' lives, and I just thought I ought to tell you."*

—LYNN E. TRIMPE, AMATYC Teaching Excellence Award Winner, "Opportunities for Excellence: Professionalism and the Two-Year College Mathematics Faculty," AMATYC, 2001, p. 33.

Let's consider an example of EQ at work in a math department. It has long been my belief that if ever students rebel again on college campuses, as they did in the 60s, it would be over the issue of colleges taking so much tuition money from them and giving so little back. Faced with large classes and/or inadequate teaching assistants instead of full-time instructors, the math department at the University of Rochester reached a crisis in 1995.

Cost-cutting measures included reducing the size of the permanent faculty in mathematics by more than *half* and the *elimination of its graduate program*. Needless to say, this caused a great upheaval in the math community. By 1996 the graduate program had been reinstated, though on a smaller scale. Less severe long-term cuts were made in the faculty. The math department salvaged its situation by applying principles of good EQ.

The leadership for the change was supplied by its chairman Douglas C. Ravenel who writes, "It is the responsibility of the mathematicians to be at least as knowledgeable as the engineers about all relevant education questions. For example, ignorance or unwillingness to master software is not a sound pedagogical argument against using it." Ravenel and his department instituted many changes, which redeemed their calculus program and the respect of the rest of the university. Among them were reaching out to all departments with math requirements, especially engineering, in an effort to learn about other departments' needs, the creation of an improved math placement test, and development of a WebWork software package that delivered grades and homework over the Internet.

Ravenel closes his article with several very pointed remarks. "Math profs cannot be merely researchers reluctantly teaching on the side in order to pick up a paycheck. In general, mathematicians tend to be too wrapped up in their research to care about their institution. Ultimately, it is the undergraduates who pay most of the bills. They need to get more back for what they put in." I am proud to add that many similar EQ-type measures have been instituted at Indiana University–Purdue University, my home campus, with the leadership of Ben Boukai, chairman of the Department of Mathematical Sciences, and David Stocum, Dean of the School of Science.

Stocum gives his views on a new university for the twenty-first century:

**1.** It should be oriented toward student learning and committed to high-quality and relevant instruction for a diverse population of undergraduates, graduates, and professional students.
**2.** It should conduct cutting-edge basic and applied research that has linkages to the undergraduate experience.

## COPING WITH COMPLETION RATES

The topic I hear discussed the most among professors at community colleges is the issue of course completion rates. It seems a foregone conclusion that a completion rate of 50% to 60% is totally unacceptable, that it should be closer to 90%. Implied is the idea that everyone can learn mathematics well. I wish that were true of my attempt to learn the clarinet as a child, or golf as an adult. I disagree with the conclusion that these completion rates should be deemed unacceptable.

Part of the pressure for high completion rates comes internally from the sheer dedication to the cause of mathematics. Neptune says, "Community colleges have been described as the most teaching-intensive component of American post secondary education. While faculty at many four-year institutions concentrate on research in their disciplines, faculty at two-year

REFERENCE
Ravenel, D. C., "Rochester Four Years Later: From Crisis to Opportunity," Notices of the AMS, September 1999, pp. 861–863.

REFERENCE
Stocum, D. A., "The Evolution of Twenty-First Century Public Education: The Urban University as a Prototype," *Metropolitan Universities Journal*, Vol. 12, No. 2, pp. 10–19, 2001.

REFERENCE
Neptune, C., *Opportunities for Excellence: Professionalism and the Two-Year College Mathematics Faculty*, AMATYC, 2001, p. 13.

institutions direct their efforts toward teaching and toward classroom research." Another part of the pressure comes from administration. Deans and department chairmen inherit that pressure from boards of trustees and state education commissions.

There is a very simple way to raise completion rates, and that is by lowering standards. In fact, I believe such actions are the main reasons enrollments in community college developmental courses have skyrocketed in recent years. The next time you teach elementary algebra, spend the whole course on solving linear equations and never tackle any word problems. You will see a dynamic increase in completion rates and, for the moment, you will look good to your fellow faculty and the dean.

Of course, the end result of lowering standards will be a total disaster when your "successful" students fail miserably in the next class. Then the department chairman and the dean will be checking on your now-questionable performance.

It is my heartfelt conviction that completion rates of 50% to 60% are excellent when we delve deeper into the situation. High school programs have many outstanding advanced placement programs to accommodate the top-notch students. I doubt these students attend community colleges.

Of those students not taking advanced placement, many take a normal sequence of math courses, completing their algebra II and/or college algebra courses in high school, and do not need remediation. They enter precalculus or calculus when they go to college.

Now what is left? The remaining students' abilities range from general mathematics, which is really seventh- and eighth-grade arithmetic taken over and over again in high school, to those with some exposure to elementary algebra, plus some students who are going back to college after a period of time away from education. The average age of community college students is about 27. Is it reasonable to expect those students to excel in math at the community college? In this context, a 50% to 60% completion is reasonable and quite good. Instructors should consider themselves successful when their students complete courses at these rates.

Here is another perspective on completion rates. The NCAA keeps careful statistics on how their athletes compare to their fellow students. Table 1 compares the graduation rates for all students to those of athletes for the 1995–96 college year. Perhaps it is not a sound comparison, but the 50% to

**Table 1**    Student vs. Student-Athlete Completion Rates

| Freshman Cohort Graduation Rates | All Students | Student Athletes |
|---|---|---|
| 1995–96 Graduation Rates | 58% | 60% |
| Four-Class Average | 57% | 59% |

*Source: NCAA.*

60% completion rates of developmental math courses are not too far from the graduation rates of athletes.

## A Better Way to Evaluate Our Success

I would like to see us adopt the following slogan to guide evaluations of success in the teaching of developmental courses:

*Think not of the quantity of students who get through these courses, think of the quality of the students who get through these courses!*

In this context, we should almost ignore completion rates. Instead, ask those who got As or Bs how successful were they in the next course or in a major that required mathematics? I once tracked our intermediate algebra students at IUPUI in such a study. Our business department wanted students to be able to take a brief calculus course with intermediate algebra as its prerequisite. Our developmental courses awarded only grades of A, B, C, F, no Ds, and I found that no student with a grade of C or below ever passed brief calculus, but those with As or Bs were quite successful.

REFERENCE

McKenney, J., and Williams, D., "A Study of Success Rates and Mathematics Backgrounds of Developmental Mathematics Students," *The AMATYC Review*, Vol. 22, No. 2, pp. 37–46, 2001.

McKenney and Williams did an excellent twenty-year tracking study of developmental math students at Northern Kentucky University. One major finding of this study was that successful completion of the Kentucky pre-college curriculum (algebra I, algebra II, and geometry) in high school did not guarantee placement into college-level mathematics courses. However, failure to complete the precollege curriculum assured the student of placement into developmental mathematics courses. Students placed into the lowest-level developmental course were very unlikely to ever pass a general studies mathematics course.

My conclusions are that as college mathematics educators, we should

- continue our all-out efforts to provide the best instruction possible to our developmental students,
- pay more attention to the quality of our students than to the quantity that pass a developmental course, and
- be more accepting about 50% to 60% completion rates. They are actually quite good when kept in perspective.

## BEING A COLLEGE MATHEMATICS EDUCATOR

Typically people with a Ph.D. in math education become college professors in education and/or math departments, teaching elementary and secondary methods, supervising secondary math majors, and doing research, if any, that relates to K–12 instruction. Some become city or state supervisors of K–12 mathematics.

It was my conviction when I worked on my Ph.D. in mathematics education that there was also much to be done in the way of math education strictly at the college level. This conviction proved to be true at IUPUI with the proliferation of developmental students and my writing textbooks in that area. There are as many problems in math education to work on at the college level as at the K–12 level.

I think every two-year and four-year college should have at least one trained professor who specializes in college mathematics education. The duties of such a professor could include:

1. Being a course coordinator of developmental, precalculus, and calculus courses, which have large enrollments
2. Conducting experiments in the area of developmental mathematics, which can be brief or elaborate in terms of statistical complexity
3. Displaying an exemplary record of teaching excellence that includes being a role model for teaching excellence and perhaps even assisting research professors in improving their teaching skills
4. Writing textbooks in the area of developmental or precalculus courses
5. Serving as a viable liaison with departments outside of mathematics to be certain that the mathematical needs of their majors are met

Accordingly, this type of professor would be rewarded in terms of salary and promotion for excellence in these areas without being required to do research in pure mathematics. Typically, a research professor is expected to excel at research, teaching, and some type of university service, and he is promoted on the basis of excellence in two of these areas. What happens, in fact, is that publishing research papers in pure mathematics becomes a necessity.

A departmental math education professor can possibly get caught in the wrong type of evaluation. That position should be evaluated according to the five tasks noted above. Let no one think I am against pure math research, I just think there should be a better balance in math departments between quality education of students and research.

It is interesting how often people think erroneously that I am some kind of mathematical research genius because I write math textbooks. I am not trained to be a research mathematician, though as this book has clarified I have done many kinds of very limited mathematics research over the years. However, the results were more in the way of minor applications than pure mathematics. I just followed some of my interests in applications.

### College Math Education Research

There is a great need for statistical math education research at the college level. It could be elaborate demanding statistics with "big $n$s", or it could be rather informal and similar to the studies I did on mastery learning and graphing calculators. I would like to see more research in the following areas:

- Mastery learning
- Tracking

- Daily quizzing
- Distance learning
- Emotional Intelligence
- Use of cumulative tests instead of chapter tests throughout a math course
- Uses of educational technology (graphing calculators, Web-enhanced courses, and software)
- Use of applications to enhance the completion rates of developmental math students.

Regarding the last item, I would especially like to know whether the emphasis on applications only yields an emotional reaction, allowing students to feel better about their math courses, or does it enable more students to succeed? (If you are considering a Ph.D. in College Math Ed, you might consider one of these as a research topic.)

## INTUITION IN MATHEMATICS EDUCATION

Over the years, I have often heard the notion of "intuition" bandied about in statements like, "That book is not very intuitive," or "Intuitively, I think that statement is true." At one time I sought a definition that I could use effectively when I taught and when I wrote.

REFERENCES
*The American Heritage Dictionary of the English Language*, 4th ed. Boston, Houghton Mifflin, 2003, p. 728.

According to *The American Heritage Dictionary of the English Language*, *intuition* is defined as: "1. The act or faculty of knowing or sensing without the use of rational processes; immediate cognition. Knowledge gained by the use of this faculty; a perceptive insight. 2. A sense of something not evident or deducible; an impression." *Intuitive* is defined as follows: "1. Of, relating to, or arising from intuition. 2. Known or perceived through intuition. 3. Possessing or demonstrating intuition."

REFERENCES
Wilder, R. L. , "The Role of Intuition," *Science*, Vol. 156, No. 3775, 1967, pp. 605–610. Wilder was president of the Mathematical Association of America from 1965 to 1966.

While these definitions gave me insight, I still sought a way to utilize intuition in my teaching and my writing of mathematics. The following is the most practical definition of intuition to achieve my goal: ". . . my teachers associated it, in some way, with experience—mathematical experience, to be more precise—and that the more experienced the mathematician became, the more reliable did his intuition become."

Thus, to teach or write intuitively, the instructor must feed the mathematical experience of students or readers. Here are some examples.

### The Commutative Law

Students in elementary algebra are studying the integers and have learned to add and subtract negative numbers. A teacher wants to get across the fact that we can add in any order; that is, teach *the commutative law*.

METHOD 1.   The teacher writes down the following formal statement:

For any real numbers $a$ and $b$, $a + b = b + a$.

Teaching this concept in such a formal way could be overpowering to many students. To feed them with mathematical experience, to teach more intuitively, consider the following methods.

METHOD 2.   The teacher has students complete the following table and look for a pattern.

| Sum | Result | Sum | Result |
|---|---|---|---|
| $5 + 3$ | 8 | $3 + 5$ | 8 |
| $-4 + 7$ | | $7 + (-4)$ | |
| $1 + (-6)$ | | $-6 + 1$ | |
| $-2 + (-10)$ | | $-10 + (-2)$ | |

METHOD 3.   The teacher shows some additions on a number line. Either way we get the same result.

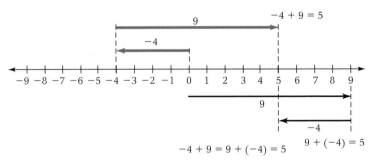

Methods 2 and 3 are more experienced based. When, if ever, the teacher or writer gets to Method 1, the student has a base of experience to understand it better. This student has been taught more intuitively.

## Intuition and Problem Solving

The better the base of experience, that is, the better the intuition in the teaching, the better the student is at problem solving, whether it be problems in a book or out in the workplace. Let's consider an applied problem from my *Introductory Algebra* book. A five-step problem-solving strategy is proposed. The steps are as follows.

The Familiarize step could also be called the Intuition step because it is there that the student is fed mathematical experience regarding the

---

*A student is taught* intuitively *when he or she is taught from a base of mathematical experience.*

REFERENCE

Bittinger, M. L., *Introductory Algebra*, 9th ed. Boston, Pearson/Addison-Wesley, 2003, p. 616.

---

### Five Steps for Problem Solving in Algebra

1. *Familiarize* yourself with the problem situation. This is the intuition step.
2. *Translate* the problem to an equation.
3. *Solve* the equation.
4. *Check* the answer in the original problem.
5. *State* the answer to the problem clearly.

---

problem. The plan is to teach enough of these experiences so that the student will feed his intuition as he does problems on his own. Let's look at a problem.

*Imax Movie Prices.* There were 270 people at a recent showing of the IMAX 3D movie *Antarctica*. Admission was $8.00 each for adults and $4.75 each for children, and receipts totaled $2088.50. How many adults and how many children attended?

There are many ways in which to familiarize ourselves with a problem situation (step 1 in the five-step problem-solving strategy). This time, let's make a guess and do some calculations. The total number of people at the movie was 270, so we choose numbers that total 270. Let's try

220 adults and

50 children.

How much money was taken in? The problem says that adults paid $8.00 each, so the total amount of money collected from the adults was

220($8),  or  $1760.

Children paid $4.75 each, so the total amount of money collected from the children was

50($4.75),  or  $237.50.

This makes the total receipts $1760 + $237.50, or $1997.50.

Our guess is not the answer to the problem because the total taken in, according to the problem, was $2088.50. If we were to continue guessing, we would need to add more adults and fewer children, since our first guess gave us an amount of total receipts that was lower than $2088.50. The steps we have used to see if our guesses are correct help us to understand the actual steps involved in solving the problem.

Let's list the information in a table. That usually helps in the familiarization process. We let $a$ = the number of adults and $c$ = the number of children.

|  | Adults | Children | Total |
|---|---|---|---|
| Admission | $8.00 | $4.75 |  |
| Number Attending | $a$ | $c$ | 270 |
| Money Taken In | 8:00$a$ | 4.75$c$ | $2088.50 |

$\longrightarrow a + c = 270$

$\longrightarrow 8.00a + 4.75c$
$\quad = 2088.50$

In step 2, translate, the student is taught to find a system of equations that fits the problem. In this case, the system is

$$a + c = 270,$$

$$8.00a + 4.75c = 2088.50.$$

But how many times do students try to race to this translation without the proper base of mathematical experience? In this case the guessing and, then the table provided this base, that is, the intuition.

We need to improve the problem-solving skills of students, teach more intuitively, that is, teach from an experience base. It takes more space in books, more time in the classroom, and more time on the part of the students, but if they are willing to spend the time, they will improve their problem-solving skills!

Let's consider one more illustration of an intuitive approach with a calculus problem on maximizing volume.

*Maximizing Volume.* From a thin piece of cardboard 8 in. by 8 in., square corners are cut out so that the sides can be folded to make a box. What dimensions will yield a box of maximum volume? What is the maximum volume?

REFERENCE
Bittinger, M. L., *Calculus and Its Applications*, 8th ed. Boston, Pearson/Addison-Wesley, 2004, pp. 251–253.

Typically in calculus, we translate the problem to a function, then use calculus techniques to find the maximum value. But to enhance our experience base after the translation, we might look at the graph of the function and make a table of function values. This could be done by hand or by using a graphing calculator or even a spreadsheet.

## Functions and Intuition

The notion of "function" has evolved over the history of mathematics. Probably no concept in elementary mathematics needs intuitive teaching as much as that of a function. Let's look at some definitions.

---

**DEFINITION 1  A Function as a Rule**

A function $f$ from a set $A$ to a set $B$ is a rule which assigns to each element of $A$, called the domain, a unique element in a set $B$. The set of all such elements in $B$ is called the range.

---

Often we have a formula that describes a function. For example,

$$f(x) = 3^x + \frac{1}{x}$$

describes the rule, pick a number in the domain, say 2, then

$$f(2) = 3^2 + \frac{1}{2} = 9\tfrac{1}{2}$$

tells us that the unique number assigned to 2 is $9\tfrac{1}{2}$. Since we can't take the reciprocal of 0, the number 0 cannot be in the domain of the function. The formula gives us a recipe for the function. Usually, the domain is understood to be all the real numbers which can be substituted. In this case, the domain is the set of all nonzero real numbers.

INTERPRETATION 1. *A Function as a Machine.* Consider the function $f$ given by the formula $f(x) = 2x + 3$. Think of $f(4) = 11$ as putting a member of the domain (an input) into the machine. The machine knows the formula $f(x) = 2x + 3$, multiplies 4 by 2, adds 3, and gives out a unique member of the range (the output), 11.

| Function: $f(x) = 2x + 3$ | |
| --- | --- |
| Input | Output |
| 4 | 11 |
| −5 | −7 |
| 0 | 3 |
| $a$ | $2a + 3$ |
| $a + h$ | $2(a + h) + 3$ |

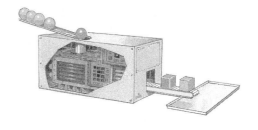

The notion of a function as a rule is encountered most in physics, chemistry, and other applied sciences.

> **DEFINITION 2   Function as a Correspondence**
> A function from a set $A$ to a set $B$ is a correspondence which assigns to each element in set $A$ a unique element in set $B$.

Each of the following is a correspondence. The first two are functions. The third is not a function.

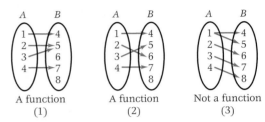

> A function   A function   Not a function
> (1)          (2)          (3)

Each arrowed connection in a correspondence can be thought of as an ordered pair. For example in (1) from $1 \to 4$ we get the ordered pair $(1, 4)$. From the ordered pairs, we can form graphs of the previous correspondences.

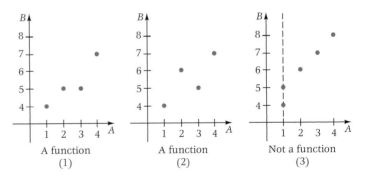

> A function   A function   Not a function
> (1)          (2)          (3)

The sets of ordered pairs of correspondences (1) and (2) are functions. For both (1) and (2) no vertical line crosses the graph more than once. But for (3) there is a vertical line that crosses the graph more than once. Thus, we could define a function as a set of ordered pairs for which no two distinct pairs have the same first coordinate.

A *relation* from $A$ to $B$ is a subset of the set of all ordered pairs $(a,b)$ where $a$ is an element of $A$ and $b$ is an element of $B$.

> **DEFINITION 3   Function as a Relation**
> A function $f$ from set $A$ to set $B$, denoted $f: A \to B$ is a relation $f$ from $A$ to $B$ such that
> **a.**   the domain of $f = A$, and
> **b.**   for every element $a$ in $A$, and every two elements $b_1$, $b_2$ in $B$, if $(a,b_1)$ is in $f$ and $(a,b_2)$ is in $f$, then $b_1 = b_2$.

A function can then be defined as a special kind of relation.

We also say that if $(a,b)$ is in $f$, then $f(a) = b$. Thus, we could rewrite part (b) of Definition 3 in two other ways.

- For every $a_1$ and $a_2$ in $f$, if $a_1 = a_2$, then $f(a_1) = f(a_2)$, or
- For every $a_1$ and $a_2$ in $f$, if $f(a_1) \neq f(a_2)$, then $a_1 \neq a_2$.

How do you think intermediate algebra students would grasp Definition 3? Not too well. Definitions 1 and 2 and Interpretation 1 provide the mathematical experience, the intuition to understand Definition 3. Now if we had 100 mathematicians in a room, they might argue over the different definitions and the interpretation, as to which is the proper way to present the notion of function. But if you wanted to present the topic to students, the use of all the definitions and the interpretation would better enable their understanding. In short, it is not a matter of which to use; use them all if eventually students can understand them.

If you are a teacher, keep in mind that improving your use of intuitive thinking, will improve its use by students.

*Remember, mathematical experience feeds intuition!*

# Epilogue

You have learned of my passion for mathematics, writing, personal relationships, and hobbies. My experience is that most students think a math textbook author lives in some kind of nerdy isolation with no dreams or adventure in his soul. Nothing could be further from the truth. There is indeed adventure in striving to create and present mathematics in a meaningful manner in the hope the mathematics will help others to fulfill their dreams. I have other dreams: some have been fulfilled; others remain in the future. I close this book recapping these dreams and sharing those for the future.

Author George Plimpton was best known for placing himself into the arenas of professional sports and writing about his experiences. From his experience as an NFL quarterback, he wrote the best-selling *Paper Lion*; from being on the professional golf tour, he wrote *The Bogey Man*; from boxing with Archie Moore, he wrote *Shadow Box;* and from pitching to major-league baseball players, he wrote *Out of My League.* He also ran with the bulls in Spain with Ernest Hemingway, played tennis with Pancho Gonzalez, and appeared in several movies. As I consider my dreams, I realize that I have led a sort of Plimpton-like existence, but lived out my dreams in a different way amid God's paradox.

As a child, I dreamed of becoming a pilot, but my eyesight prevented it. Instead, God blessed me one day with the opportunity to teach at the United States Air Force Academy, where I rode in a jet trainer and a glider. More important, my books enabled thousands of U.S. pilots to learn mathematics in order that their dreams might be fulfilled.

I dreamed of being a professional bowler and lived out that dream by earning membership in the Professional Bowlers Association, and bowling a 297 game in a national tournament. Now I can walk in the circles of professional bowlers, but I can't compete with them. But, I was able to use mathematics to develop a more effective procedure for making spares and to reveal an undiscovered result regarding the variance of the scores of professional bowlers.

I dreamed of being a major-league baseball player only to have that dream shattered also by poor eyesight. In 1954 my dream of seeing the

Cleveland Indians win the World Series was shattered when New York Giant Willie Mays made the most incredible catch in baseball history off the bat of my hero, Cleveland Indian Vic Wertz. God gave my baseball dream back to me when I wrote a book on hitting in 1993 with then Giants manager, Dusty Baker. Dusty introduced me to Willie Mays at a banquet honoring Dusty with the National Manager of the Year Award. Through our book *You Can Teach Hitting*, thousands of players and coaches improved their skills at hitting or the teaching of hitting. I also made a mathematical discovery regarding two-seam fastballs versus four-seam fastballs: the two-seam pitch arrives at the plate 30 inches behind the four-seam pitch. Dave Wallace, current pitching coach of the Boston Red Sox, and other major league coaches were ecstatic to learn this result.

My obsession with western movies fed my passion for the beauty of the West. I fulfill that dream with frequent trips to Utah to hike in Arches and Canyonlands National Parks with my two sons Lowell and Chris Bittinger. That scenery takes me as close to God and heaven as I can be on this earth.

You might wonder whether my math textbook writing was another dream that God fulfilled. While it would be glamorous to write that I yearned, dreamed, and pined to be an author, in truth such a profession was far beyond my finite dreams. It was as if God opened a door and I shot through it on a space shuttle. God had an infinitely bigger dream than my mind could imagine! As someone once said, "Life is what happens while you're making other plans."

So what's still on the list? Corny as it sounds, I dream of meeting my favorite hero of western movies, Monte Hale. He is 84 as I write this, and I see him from time to time on documentaries about western movies. He appeared in the 1956 movie, *Giant*, one of my all-time favorites. I fantasize about jumping on a horse, my saddle bags full of math books, and riding over a hill into a small town; then by handing out the math books, I resolve all their learning problems. Yeah, weird! But, these are the workings of my heart.

Regarding baseball, I would like to stay healthy enough to continue going to adult baseball fantasy camp every year for the rest of my life. I also dream of seeing at least one baseball game in every major-league city before I die. At moments like these, I fantasize about trading my math textbook home runs for major league home runs, but that can only happen in heaven.

Regarding other sports dreams, I now am able to make about 1 good golf shot out of 5. That makes me a bad golfer. I'd like to make 3 good shots out of 5. That would certainly make me a better golfer, but I suspect that God will continue to use this game to keep me humble. My dream about bowling is to roll that elusive 300 game to cap the joy and frustration of games of 297 and 299.

My father took me often as a child to a greasy hamburger restaurant in Akron, Ohio, called Thacker's. We sat at the counter—there were no tables. Behind the counter was a cook who took orders as well. He featured a tooth-pick in his mouth and an apron covered with grease, mustard, and other grime accumulated from wiping his hands. Served with only the condiments of pickle, onion, and mustard (no catsup), the "hamburgs" cost 15¢ and were so small that you had to order five just to get full. I always had a grape soda and a piece of cherry pie to fill the cracks. It was food heaven. I would like to wear such an apron and fry "hamburgs" for two weeks in such a restaurant. Just think of the joy of interacting with my customers about mathematics and the philosophy of life. The closest place I have encountered since was a place in Moab, Utah, called the "Westerner Grill," where the locals came to drink coffee and philosophize. It exists no more.

*Westerner Grill*

As a truck driver for Standard Oil of Ohio, my father delivered gasoline to farmers surrounding Akron. Once in awhile he would stop home in the late afternoon and take me back to work with him in that truck. It was big and noisy and somewhat intimidating to a young child. I'd like to learn how to drive a big rig, and wheel it all over the West for two weeks. It would be a tribute to my Dad, a hard-working man who could have finished college had his father not forced him to leave high school after his sophomore year because "It's time to get a job and make some money." I thank God that greater wisdom prevailed in the mind of my grandmother, who raised me. Otherwise, I would not have a college education, let alone be writing books.

Regarding my family, I am very proud of my sons Lowell and Chris and the way they care for their wives and children and embrace their professions. I dream of attending the wedding of each of my granddaughters and seeing them fulfill God's dreams for their lives. With Elaine, I dream of more traveling, cherishing our grandchildren, and finishing our lives together in spiritual growth, somehow embracing the trials of death in a gracious manner.

My dreams as a mathematics educator are twofold. The first is that the need for developmental (remedial) mathematics at the college level would virtually disappear. Remember, these are college students taking junior-high and high school subjects. I know those who have been away from college for a period of time will always need brush-up courses, but it would be great if the rest came to college having satisfied their math prerequisites. Then what would I write? Were I not retired, I might just work on my brief calculus book or college algebra and trigonometry.

My second dream as a mathematics educator is to teach a math class in which each student satisfies the following criteria.

The *ideal student*

1. has studied and mastered all the prerequisites for the course.
2. loves math and cannot wait to have the mystery of a new math course revealed.

3. studies one section ahead every day.
4. does the homework each night so as not to get behind.
5. gets to class everyday, on time, and sits up front to see and hear effectively.
6. does some exercises beyond the assignment to prepare to take tests that have no answers provided.
7. is a well-read, well-rounded individual—in touch with literature, science, and philosophy—and is not afraid to ask questions that might connect these subjects to mathematics.

By the way, "seven" in Hebrew implies perfection. If this dream came true, it would be an **MPX** of the highest order.

**MPX** MUSIC: "The Magnificent Seven," by Elmer Bernstein, arranged by Christopher Palmer, track 3 on the CD *Round-Up*, performed by Erich Kunzel and the Cincinnati Pops Orchestra, 1986, TELARC.
MOVIE: *The Magnificent Seven* (1960) with Yul Brynner, Steve McQueen, Charles Bronson, Robert Vaughn, James Coburn, et al. Perhaps the "bad guys" in this movie represent the bad ways of studying mathematics, which get wiped out.

*Prov 2:10 For wisdom will fill your heart, and knowledge will fill you with joy.*
—NEW LIVING TRANSLATION

**REFERENCES**
Eldredge, J., *Wild at Heart*. Nashville, Thomas Nelson Publishers, 2001.
Eldredge, J., *Waking the Dead*. Nashville, Thomas Nelson Publishers, 2003.

Jer 29:11 *"For I know the plans I have for you," says the Lord. "They are for good and not for disaster, to give you a future and a hope."*
—NEW LIVING TRANSLATION

From a philosophical standpoint I have two dreams. The first involves *wisdom* and the second *paradox*. I define *wisdom* as the acquisition and application of knowledge. A key to finding wisdom is questioning. Questioning makes us analyze, and when we analyze we find insight. The challenge is to apply that insight in my life.

At this point in my life, realizing that the road is precarious and difficult to find, I absolutely crave *wisdom*. I just can't get enough. This probably stems from the realization that I have only a short time to live and had best use my precious time to attain all the wisdom I can. Following my heart attack in 1998, I developed a passionate quest for theological wisdom, which I pursue through the Bible and books by authors like M. Scott Peck, Philip Yancey, Hugh Ross, and John Eldredge. If a book does not soon supply me with fresh bits of wisdom, I move on to another.

John Eldredge wrote two of my favorite theological books, *Wild at Heart* and *Waking the Dead*. The wisdom I drew from Eldredge's books is that God plants in each person's heart a wonderful, fulfilling plan, if only the person will pursue it with passion. This idea ended a long struggle that started in my youth. My fundamental upbringing had deeply rooted the idea that the only way I could serve God was in full-time Christian work such as being a minister or a missionary—prospects that did not excite me. Then I learned that God does not give us a plan that we would dread. He gives us something we will thrive at doing. Accordingly, that plan can involve being a grocery clerk, a salesclerk, a mail carrier, a teacher, a doctor, a mathematician, or an author.

My second philosophical dream is to embrace *paradox*. A *paradox* is a statement of two parts that seems contradictory, but may still be true. For

*Don't ask yourself what the world needs; ask yourself what makes you come alive. And then go and do that. Because what the world needs is people who have come alive.*

—HAROLD WHITMAN

REFERENCE

Peck, M. S., *Golf and the Spirit.* New York, Three Rivers Press, 1999, p. 114.

me, a paradox can be a statement that expresses opposing statements or viewpoints. In other situations, a paradox expresses two sides of an issue. An example of a paradox in everyday life is, "What would life be like if there were no hypothetical questions?"

I see paradox throughout my life and dreams. So many times I sense that life is positive (good), arranged just the way I want it, only to have it rocked by something negative (bad). I often "write math books in my mind." I recall one day that a man held a door for me and, being lost in my thoughts concerning my writing, I went right through it without thanking him. He then ripped me verbally. I deserved it! Two sides of the paradox came through, the 'good' in my goal of writing well and the 'bad' in the kindness I failed to express to a fellow man.

Another paradox occurs in letters from my students. For example, one tells me how much my books have enhanced her learning. Another compares my writing to gutter trash and calls me a "math geek."

Another form of paradox occurs to me every time I use proof by contradiction in mathematics. It amazes me that to prove $S$, I might first assume *(not S),* and try to find a sentence $P$ such that both $P$ and *(not P)* are true at the same time. In mathematics, the statement *P and (not P)* is always false by the rules of mathematical logic, and is called a *contradiction.* A simple example is "$x = 2$ and $x \neq 2$."

How can truth come out of paradox? In mathematics lots of truth comes out of proof by contradiction. In life, lots of truth comes out of paradox. I learned this bit of wisdom from M. Scott Peck, who asserts that "paradox is the foundation of all essential truth, and the capacity to embrace paradox is the key to psycho-spiritual growth, whether it be on the golf course, in the boardroom, over the stove, or even in the bedroom." I want psycho-spiritual growth, so I am willing to embrace paradox to find it.

When I sense both 'good' and "not good," I think of paradox. I believe that God's infinite potential can resolve paradox. The most profound example of God resolving a paradox in my life followed my heart attack. After stents had been used to open my clogged arteries, I began a very aggressive diet and exercise program to prevent further heart problems. The 'good' was the false sense of security before my heart attack; I had passed a treadmill test with flying colors just five weeks earlier. The 'not good' was the realization that I was indeed mortal.

The wisdom I found is that God is able to resolve a paradox; in this case the resolution was twofold. First and foremost, I am in better physical health now than ever in my life. A kind of empowerment comes with exercise. I can do activities like sports and hiking with high endurance. Not only am I more heart-healthy, but exercise strengthens the body's immune system and inhibits heart disease, diabetes, and numerous other health problems.

The second way God resolved the paradox came quite unexpectedly. Running on a treadmill is monotonous and appeared to be such a waste of "mental" time. Then I saw a few people reading and decided to give it a try. Although some people have trouble reading on the treadmill, I have the advantage of only one good eye—I can focus on the words. In the past five years I have read 50 to 100 books and stacks of magazines and articles printed off the Internet. I read some math books, but also the Bible, and numerous theological, scientific, and other nonfiction books. Needless to say, the wisdom gained has been quite satisfying. I would venture to say that without all that reading, this book would not have been possible. God did resolve the paradox. In the process I think I grew psycho-spiritually.

My final dream is to publish the companion volume to this book entitled *Mathematics and the Spirit*, in which I develop the notion of combining my faith with my mathematics. Historically, Greek philosophers Plato and Aristotle were credited with the marriage of science and religion in the eyes of mankind. Mathematicians like Descartes, Pascal, Newton, Euler, Cauchy, Russell, and Whitehead, though not all theists, were philosophers as well. It is my faith axiom that every mathematician (or scientist) should return to a union or integration of mathematics (science) and theology in their lives. I defend this position in *Mathematics and the Spirit* using mathematics as a framework or model for spiritual insight into the following topics:

- Probability and prophecy
- Modeling the growth of Christian evangelism
- The power of prayer
- Higher dimensions
- Paradox and psycho-spiritual growth
- Faith axioms

The first three topics are more analytic. The last three are more philosophical, combining ideas from science, mathematics, and theology. More information and greater depth can be found on my Web site www.marvbittinger. com or in the book *Mathematics and the Spirit*.

Do my dreams include retirement? I am intrigued by how often friends or slight acquaintances ask me now if I have retired or plan to do so. It makes me feel old and humbles me to think that I still might look young from all the exercise I do for my health. In truth, my mind thinks it is young even while my body is aging.

A statement by my hero M. Scott Peck applies, "It is the issue of *power*. . . . Since power confers status, nothing can be more enhancing of self-esteem." In my case, the word *power* is a paradox. One side of the paradox is enjoying the benefits of all the hard work that textbook writing entails. Nothing has fed my self-esteem more than my writing. I was unable to be an athlete or a pilot. But I could be a writer, and I imagine there are a number of athletes and pilots who wished they could be writers. On the other side, I seem to possess less of the passion a writer so desperately must have. Yet, it is difficult to retire from writing. Peck also states that "power in all its varieties can

REFERENCE

Peck, M. S., *In Search of Stones*. New York, Hyperion, 1995, pp. 408–411.

be so alluring, it is no wonder that most people cling to it for all it is worth and as long as they can—usually until it is wrested from them by a palace coup, a changing of the guard, a disgrace, a debilitating illness, death, or that relatively recent invention of society, 'mandatory retirement.'" My "palace coup" might be that my publisher no longer wants to publish my work or that my co-authors are honest or subtle enough to tell me that I am more trouble than I am worth.

There is another paradox in the retirement issue. First, Peck savors the notion that "There is also the power to serve." In that sense I will probably never retire as long as I can serve my students by my writing, my God by using the talents He blessed me with, my co-authors by giving them advice, and my readers with a book such as this, which allows me to pass on the wisdom of my years.

Second, Max Lucado summarizes my thoughts on retirement: "Your last chapters can be your best. Your final song can be your greatest. It could be that all of your life has prepared you for a grand exit. God's oldest have always been among his choicest."

A final  for the joy of being able to complete this book.

*A man doesn't begin to attain wisdom until he realizes that he is no longer indispensable.*
—ADMIRAL BYRD

REFERENCE
Lucado, M., *Grace for the Moment.* Nashville, J. Countryman, A Division of Thomas Nelson, 2000, p. 179.

**MPX** MUSIC: "This Is the Moment," performed by the Purdue University Varsity Glee Club, track 12 on the CD *Musical Memories, Vol. II,* Purdue Musical Organizations of Purdue University, www.purdue.edu/PMO/

# Acknowledgments

This book would not have been possible without Greg Tobin, Vice President and Publisher, Addison-Wesley Publishing Company, who envisioned it, and Charles D. Taylor, whose diligent editing provided the polishing I so desperately needed.

To all my reviewers, thank you for pertinent criticisms provided against the background of being close friends. They are Jim Biddle, Matt Hassett, Ed Zeidman, Dave Neuhouser, and Sybil MacBeth.

To all my past and present editors, I praise you for all the hard work behind the scenes that allows us as authors to build better books. Among them are Chuck Taylor, Liz Hacking, Pat Mallion, Stu Johnson, Nancy Kralowitz, Betsy Burr, Jason Jordan, Greg Tobin, Jennifer Crum, Steve Quigley, Dave Geggis, and Bill Hoffman.

To fellow authors who granted me interviews, I thank them for altering busy schedules to allow me to reveal more about the creative aspect of writing. They are M. Scott Peck, Philip Yancey, Charles D. Taylor, Phillip R. Craig, and Neil A. Campbell.

To all my encouragers, I appreciate the kind words of support and interest, as well as the research help. They are Elaine S. Bittinger, Christopher N. Bittinger, Tricia A. Bittinger, Lowell D. Bittinger, Karen E. Bittinger, Benzion Boukai, Rose Beetem, Sally Lympany, Carole and Larry Mattler, Richard and Lorraine Bonewitz, David Rodriguez, and all my spiritual compatriots in our Wednesday theology book group.

To all my production editors, who labor so hard to produce quality books but are so seldom recognized. Nobody does designs better than Geri Davis. She proved it again on this book. Though she did not copy edit this book, Martha Morong has been a loyal, thorough compatriot throughout numerous revisions of my books. Joan Flaherty was the copy editor on this book, and she helped me rewrite it better. Jenny Bagdigian, Kathy Manley, and Karen Guardino coordinated production to a tee. Karen, who retired during the production of this book, has been outstanding overseeing the tasks of many of my books over the years. RoseAnne Johnson was superb as project manager.

Barbara Johnson and Mike Rosenborg did an exacting check of the math in the manuscript.

Many professional baseball people have become friends and supporters and directly or indirectly provided information that led to my ability to discover mathematics amid the green grass, chalklines, and red brick dust. Many were associated with the Los Angeles Dodger Adult Baseball Camps and the San Francisco Giants Baseball Fantasy Camps. They are Dusty Baker, Wendell Kim, Bob Lillis, Tom Beyers, Dave Wallace, Reggie Smith, Ralph Branca, Carl Erskine, Vida Blue, Tommy Davis, Bob Brenly, Guy Wellman, and Clem Labine.

To my many bowling instructors, I praise you for your patience in trying to teach a passionate but unathletic person some bowling skills. They are Mike Aulby, Ed Zeidman, Charlie Garinger, Randy Stoughton, Jim Acheson, Tom Kouros, Don Johnson, Scott Mills, Ron Adkins, Robert Strickland, John Combs, and Bill Rydman.

To all the students and instructors learning or teaching from my books, you fulfilled my dream of writing mathematics textbooks that might lead to fulfilling your dreams. While it has been impossible to know you all, enough feedback has come my way that I indeed feel fulfilled. Maybe I will get acquainted with all of you in heaven?

Last but not least, I must thank Addison-Wesley Publishing Company, a division of Pearson Education, for years of support and encouragement. When I think of AW, I think of opportunity and service. They give me and thousands of authors the opportunity to be of service to students as they study from a textbook. Thank you! It's been a great ride.

# Answers

Answers to the "Pursuing Further" queries:

## CHAPTER 2

### Elaine's Problem (p. 17)

1. Consider $\frac{5}{7} = \frac{10}{14}$. The assertion would be invalid. The repetend is 7, not 14, and there aren't 14 possible remainders; nor is there a repetend of length 13.

2. Let $n = 0.999999\dots$. Then $10n = 9.999999\dots$, so $10n - n = 9n = 9$. Dividing by 9 on both sides, we find that $n = 1$.

3. They rotate, so to speak, through the sequence of six digits of the repetend, 142857, 285714, 428571, 571428, 714285, and 857142.

### Wendy's Hamburgers (p. 19)

1. $2^{12} = 4096$ kinds of pizza.
2. $3 \cdot 2 \cdot 2^{12} = 24{,}576$ kinds of pizza.
3. $26 \cdot 2^{24} = 436{,}207{,}616$.

### It's a Small, Small World (p. 22)

1. B: 1.5%; C: 3.0%; D: 3.1%; E: 13.5%; F: 1.9%; G: 2.1%; H: 4.4%; I: 5.9%; J: 0.2%; K: 0.6%; L: 4.6%; M: 2.5%; N: 6.5%; O: 7.7%; P: 2.6%; Q: 0.04%; R: 6.7%; S: 8.2%; T: 8.6%; U: 2.6%; V: 1.2%; W: 1.4%; X: 0.2%; Y: 1.4%; Z: 0.2%
2. 39%
3. 61%
4. Check out a Dvorak keyboard using an Internet search engine. The home row consists of all of the highest occurring letters. The vowels are all together. Virtually all of the least occurring letters are on the bottom row. For new typists it has been shown that DVORAK is much easier to learn. From a comfort standpoint, DVORAK stands out because most of the typing takes place on the home row.
5. **a.** T, S, N, R, L **b.** E **c.** yes
6. **a.** 82% **b.** 17% **c.** 1% **d.** About 20

7.  **a.** 83.4%  **b.** yes
8.  About 10.2%; about 7.1%

## Up, Up, and Away! (p. 29)

1.  **a.** $P(t) = 12e^{0.1t}$  **b.** 2010: $884.40; 2020: $2404  **c.** 2011; 2018
2.  Left for the reader
3.  *Beautiful Mind; Good Will Hunting; The Mirror Has Two Faces*
4.  Left to the reader
5.  Left to the reader

## The Dead Calculus Professor (p. 32)

1.  7 P.M.
2.  Left to the reader

## The Tower of Hanoi Problem (p. 39)

1.  We first list $S_n$, $S_1$, $S_k$, and $S_{k+1}$.

    $S_n$:        $1 + 3 + 5 + \cdots + (2n - 1) = n^2$
    $S_1$:        $1 = 1^2$
    $S_k$:        $1 + 3 + 5 + \cdots + (2k - 1) = k^2$
    $S_{k+1}$:     $1 + 3 + 5 + \cdots + (2k - 1) + [2(k + 1) - 1] = (k + 1)^2$

    (1) *Basis step.* $S_1$, as listed, is true.
    (2) *Induction step.* We let $k$ be any natural number. We assume $S_k$ to be
        true and try to show that it implies that $S_{k+1}$ is true. Now $S_k$ is
    $1 + 3 + 5 + \cdots + (2k - 1) = k^2$.
    Starting with the left side of $S_{k+1}$ and substituting $k^2$ for $1 + 3 + 5 + \cdots + (2k - 1)$, we have

    $$1 + 3 + \cdots + \underbrace{(2k + 1)}\ + [2(k + 1) - 1]$$
    $$\downarrow$$
    $$= k^2 + [2(k + 1) - 1] = k^2 + 2k + 1 = (k + 1)^2.$$

    We have derived $S_{k+1}$ from $S_k$. Thus we have shown that for all natural numbers $k$, $S_k \rightarrow S_{k+1}$. This completes the induction step. It and the basis step tell us that the proof is complete.

2.  The proof is in the next part of the chapter, "My Golf Problems."
3.  We first list $S_n$, $S_1$, $S_k$, and $S_{k+1}$.

    $S_n$:   $\dfrac{1}{2} + \dfrac{1}{4} + \dfrac{1}{8} + \cdots + \dfrac{1}{2^n} = \dfrac{2^{n-1}}{2^n}$

    $S_1$:   $\dfrac{1}{2} = \dfrac{2^1 - 1}{2^1}$

    $S_k$:   $\dfrac{1}{2} + \dfrac{1}{4} + \dfrac{1}{8} + \cdots + \dfrac{1}{2^k} = \dfrac{2^k - 1}{2^k}$

    $S_{k+1}$:   $\dfrac{1}{2} + \dfrac{1}{4} + \dfrac{1}{8} + \cdots + \dfrac{1}{2^k} + \dfrac{1}{2^{k+1}} = \dfrac{2^{k+1} - 1}{2^{k+1}}$

(1) *Basis step.* We show $S_1$ to be true as follows:

$$\frac{2^1 - 1}{2^1} = \frac{2 - 1}{2} = \frac{1}{2}.$$

(2) *Induction step.* We let $k$ be any natural number. We assume $S_k$ to be true and try to show that it implies that $S_{k+1}$ is true. Now $S_k$ is

$$\frac{1}{2} + \frac{1}{4} + \frac{1}{8} + \cdots + \frac{1}{2^k} = \frac{2^k - 1}{2^k}.$$

Starting with the left side of $S_{k+1}$ and substituting

$$\frac{2^k - 1}{2^k} \quad \text{for} \quad \frac{1}{2} + \frac{1}{4} + \cdots + \frac{1}{2^k},$$

we have

$$\underbrace{\frac{1}{2} + \frac{1}{4} + \frac{1}{8} + \cdots + \frac{1}{2^k}} + \frac{1}{2^{k+1}}$$

$$\downarrow$$

$$= \frac{2^k - 1}{2^k} + \frac{1}{2^{k+1}} = \frac{2^k - 1}{2^k} \cdot \frac{2}{2} + \frac{1}{2^{k+1}} = \frac{(2^k - 1) \cdot 2 + 1}{2^{k+1}}$$

$$= \frac{2^{k+1} - 2 + 1}{2^{k+1}} = \frac{2^{k+1} - 1}{2^{k+1}}.$$

We have derived $S_{k+1}$ from $S_k$. Thus we have shown that for all natural numbers $k$, $S_k \to S_{k+1}$. This completes the induction step. It and the basis step tell us that the proof is complete.

4. $S_1$: 2 is a factor of $1^2 + 1$.
   $S_k$: 2 is a factor of $k^2 + k$.
   $(k + 1)^2 + (k + 1) = k^2 + 2k + 1 + k + 1$
   $= k^2 + k + 2(k + 1)$
   By $S_k$, 2 is a factor of $k^2 + k$; hence 2 is a factor of the right-hand side, so 2 is a factor of $(k + 1)^2 + (k + 1)$.

5. $S_n$:    $|\sin (nx)| \le n|\sin x|$
   $S_1$:    $|\sin x| \le |\sin x|$
   $S_k$:    $|\sin (kx)| \le k|\sin x|$
   $S_{k+1}$:   $|\sin (k + 1)x| \le (k + 1)|\sin x|$
   (1) *Basis step:* $S_1$ is true because $|\sin x| = |\sin x|$.
   (2) *Induction step:* Assume $S_k$. Deduce $S_{k+1}$. Starting with the left side of $S_{k+1}$, we have
      $|\sin (k + 1)x|$
   $= |\sin (kx + x)|$
   $= |\sin kx \cos x + \cos kx \sin x|$
      $\le |\sin kx \cos x| + |\cos kx \sin x|$   By the triangle inequality
   $= |\sin kx|\,|\cos x| + |\cos kx|\,|\sin x|$
      $\le k|\sin x|\,|\cos x| + |\cos kx|\,|\sin x|$   By $S_k$
   $= |\sin x|(k|\cos x| + |\cos kx|)$
      $\le |\sin x|(k + 1)$   $|\cos x| \le 1$   and   $|\cos kx| \le 1$.

## My Golf Problems (p. 44)

**1.** Sum $= 1 + 3 + 6 + 10 + \cdots + \dfrac{n(n + 1)}{2}$

**2.**

$$S_n: 1 + 3 + 6 + 10 + \cdots + \frac{n(n + 1)}{2} = \frac{n(n + 1)(n + 2)}{6}$$

$$S_1: \frac{1(1 + 1)}{2} = \frac{1(1 + 1)(1 + 2)}{6}$$

$$S_k: 1 + 3 + 6 + 10 + \cdots + \frac{k(k + 1)}{2} = \frac{k(k + 1)(k + 2)}{6}$$

$$S_{k+1}: 1 + 3 + 6 + 10 + \cdots + \frac{(k + 1)(k + 2)}{2} = \frac{(k + 1)(k + 2)(k + 3)}{6}$$

(1) *Basis step:* $S_1$ is true by substitution.

(2) *Induction step:* Assume $S_k$. Deduce $S_{k+1}$.

Starting with the left side of $S_{k+1}$, we have

$$1 + 3 + 6 + 10 + \cdots + \frac{(k + 1)(k + 2)}{2}$$

$$= 1 + 3 + 6 + 10 + \cdots + \frac{k(k + 1)}{2} + \frac{(k + 1)(k + 2)}{2}$$

$$= \frac{k(k + 1)(k + 2)}{6} + \frac{(k + 1)(k + 2)}{2}$$

$$= \frac{k(k + 1)(k + 2) + 3(k + 1)(k + 2)}{6} \quad \text{by } S_k$$

$$= \frac{(k + 1)(k + 2)(k + 3)}{6}$$

**3.** $T(x) = \dfrac{x(x + 1)(x + 2)}{6} = \dfrac{1}{6}x^3 + \dfrac{1}{2}x^2 + \dfrac{1}{3}x$

**4.** $T(5) = 35$

**5.** $494.8 \text{ cm}^3$

## My Bowling Problems (p. 54)

**1.** Art: 181   Bob: 176   Carl: 158   Denny: 190   Fred: 180

## CHAPTER 3

## Baseball's Strike Zones (p. 60)

**1.** $510 \text{ in.}^2$
**2.** $334 \text{ in.}^2$
**3.** $176 \text{ in.}^2$
**4.** $34.5\%$

## Four Seams vs. Two Seams (p. 66)

1. About 31 in.
2. About 29 in.
3. Yes
4. **a.** About 0.57 sec  **b.** About 0.424 sec
   **c.** To a distance of about 34 ft from the release point. Each foot closer means roughly about 1 more mile per hour. When I informed Dusty of this notion, he said it couldn't be used in practice because it would alter the visual effect of the ball coming off the pitcher's hand. In effect, it would just be too close.

## Tale of the Tape (p. 70)

1. 421 ft

## The Lost War Years (p. 75)

1. 41, 42, 43, 44, 45   493 + 41 + 42 + 43 + 44 + 45 = 708
2. 107 – 10 + 20 + 21 + 22 + 23 + 24 + 25 + 26 + 27 + 28 + 29 + 30 + 31 = 403
   This total would have been sufficient to put him in the Hall of Fame, though typically a pitcher's totals taper off toward the end of his career.

## Baseball Statistics (p. 84)

1.

| Player | AB | H | 2B | 3B | HR | W | SF | HBP | AVG | OBP | SLG | OPS |
|---|---|---|---|---|---|---|---|---|---|---|---|---|
| Hank Aaron | 12,364 | 3771 | 624 | 98 | 755 | 1402 | 121 | 32 | .305 | .374 | .555 | .929 |
| Babe Ruth | 8399 | 2873 | 506 | 136 | 714 | 2062 | N/A | 43 | .342 | .474 | .690 | 1.164 |
| Willie Mays | 10,881 | 3283 | 523 | 140 | 660 | 1464 | 91 | 44 | .302 | .384 | .557 | .941 |
| Barry Bonds | 8725 | 2595 | 536 | 74 | 658 | 2070 | 84 | 84 | .297 | .433 | .602 | 1.035 |
| Mark McGwire | 6187 | 1626 | 252 | 6 | 583 | 1317 | 78 | 75 | .263 | .394 | .588 | .982 |

*N/A = Not available, use 0.*

2. 4.000; 0.000

3. Walsh: 1.82; Joss: 1.89; Brown: 2.06; Ward: 2.10; Mathewson: 2.13

4. Leonard: 0.96; Brown: 1.04; Gibson: 1.12; Mathewson: 1.144; Johnson: 1.145

5. **a.** 4.00; 4.50; 5.14; 6.00; 7.20; 12.00; 18.00; 54.000; 108.000  **b.** ∞  **c.** ∞

## The Formulas of Bill James (p. 95)

1. About 40%. Note that this is higher than for Bonds, even after Bonds hit 658 home runs at age 39.

2. 2.7%, much less than the probability of his breaking the home run record.

3. 2.4%

## The Magic Number (p. 98)

1. 8
2. 2
3. 1
4. 31

## CHAPTER 4

### Brain Stretchers (p. 108)

**1.** Linamint  **2.** I will if it is within my power.  **3.** Raising a piece of pie to the teeth  **4.** Yes, if your morph your mind between the symbols and the mathematics  **5. a.** 12 barn dances **b.** 15 square dances **c.** Ice cube **d.** 612 rock bands  **6.** $2L - 2L = 0 \cdot L =$ Noel  **7.** Noah's ark  **8.** Box of sand, or sandbox  **9.** Man overboard  **10.** I understand  **11.** Reading between the lines  **12.** Crossroads  **13.** Three degrees below zero  **14.** Neon light  **15.** Pair of dice, or paradise  **16.** Fractional notation  **17.** Six feet under ground  **18.** Mind over matter  **19.** Backwards glance  **20.** Mixed numerals  **21.** log Cabin + C = Houseboat  **22.** Ice cube  **23.** A Real square

### A Funny Math Test (p. 109)

**1.** y-ponents  **2.** Sine on the dotted line  **3.** A solution of right triangles  **4.** A solution of oblique triangles  **5.** (1) "errers" is mispelled; (2) "is" should be "are"; (3) "three" should be "two"  **6.** Both  **7.** All represent the empty set, $\varnothing$  **8.** They are all "square."  **9.** They are all "imaginary"  **10.** Each can be represented as the "union of mutually exclusive subsets."  **11.** All are "prime"  **12.** Most algebra students would say, "Yes" or "Right on!"  **13.** One ordered pears and the other ordered $\pi$  **14.** It means that 37 out of 10,000 died and 2 more were at the "point" of death  **15.** Eskimo $\pi$  **16.** 15  **17.** It is what you plant to get a square tree.  **18.** A drastic diet  **19.** Strontium 90, carbon 14  **20.** $9W$  **21.** Adam 812  **22.** They are both "dense."  **23.** The sons of the squaw of the hippopotamus is equal to the sons of the squaws of the other two hides.  **24.** Left to the reader.

## CHAPTER 5 (p. 119)

### Holiday Wish

**1.** $M$  **2.** $e$  **3.** $rr$  **4.** $y$  **5.** $X$  **6.** $m$  **7.** $a$  **8.** $s$